Numerical Analysis of Spectral Methods:

Theory and Applications

DAVID GOTTLIEB
Tel-Aviv University
and
STEVEN A. ORSZAG
Massachusetts Institute of Technology

SOCIETY for INDUSTRIAL and APPLIED MATHEMATICS

PHILADELPHIA, PENNSYLVANIA 19103

Printed for the Society for Industrial and Applied Mathematics by
J. W. Arrowsmith Ltd., Bristol BS3 2NT, England

Contents

Preface

Spectral methods involve seeking the solution to a differential equation in terms of a series of known, smooth functions. They have recently emerged as a viable alternative to finite difference and finite element methods for the numerical solution of partial differential equations. The key recent advance was the development of transform methods for the efficient implementation of spectral equations. Spectral methods have proved particularly useful in numerical fluid dynamics where large spectral hydrodynamics codes are now regularly used to study turbulence and transition, numerical weather prediction, and ocean dynamics. In this monograph, we discuss the formulation and analysis of spectral methods. It turns out that several features of this analysis involve interesting extensions of the classical numerical analysis of initial value problems.

This monograph is based on part of a series of lectures presented by one of us (S.A.O.) at the NSF–CBMS Regional Conference held at Old Dominion University from August 2–6, 1976. This conference was supported by the National Science Foundation.

We should like to thank our colleagues M. Deville, M. Dubiner, M. Gunzburger, B. Gustaffson, D. Haidvogel, M. Israeli, and J. Ortega for helpful discussions. We are grateful to E. Cohen, A. Patera, and K. Pitman for their assistance in preparing graphs and tables. Some calculations were performed at the Computing Facility of the National Center for Atmospheric Research which is supported by the National Science Foundation. One of us (D.G.) would like to acknowledge support by the National Aeronautics and Space Administration while in residence at ICASE, NASA Langley Research Center, Hampton, Virginia. Both authors would like to acknowledge support by the Fluid Dynamics Branch of the Office of Naval Research and the Atmospheric Sciences Section of the National Science Foundation.

Hampton, Virginia
Cambridge, Massachusetts
September 1977

SECTION 1

Introduction

In this monograph we give a mathematical analysis of spectral methods for mixed initial-boundary value problems. Spectral methods have become increasingly popular in recent years, especially since the development of fast transform methods (see § 10), with applications in numerical weather prediction, numerical simulations of turbulent flows, and other problems where high accuracy is desired for complicated solutions. We do not discuss the sophisticated applications of spectral methods here. A survey of some applications is given in § 15. Instead, we concentrate on the development of a mathematical theory that explains why spectral methods work and how well they work. Before presenting the theory, we begin by giving some simple examples of the kinds of behavior that we wish to explain.

Spectral methods involve representing the solution to a problem as a truncated series of known functions of the independent variables. We shall make this idea precise in § 2, but we can illustrate it here by the standard separation of variables solution to the mixed initial-boundary value problem for the heat equation.

EXAMPLE 1.1. *Fourier sine series solution of the heat equation.* Consider the mixed initial-boundary value problem

$$\frac{\partial u(x,t)}{\partial t}=\frac{\partial^2 u(x,t)}{\partial x^2}, \qquad 0<x<\pi, \quad t\geqq 0, \tag{1.1a}$$

$$u(0,t)=u(\pi,t)=0, \qquad t>0, \tag{1.1b}$$

$$u(x,0)=f(x), \qquad 0\leqq x\leqq \pi. \tag{1.1c}$$

The solution to (1.1) is

$$u(x,t)=\sum_{n=1}^{\infty} a_n(t)\sin nx, \tag{1.2}$$

$$a_n(t)=f_n\, e^{-n^2 t}, \qquad n=1,2,\cdots, \tag{1.3}$$

where

$$f_n=\frac{2}{\pi}\int_0^{\pi} f(x)\sin nx\, dx, \qquad n=1,2,\cdots \tag{1.4}$$

are the coefficients of the Fourier sine series expansion of $f(x)$. Recall that any function in $L_2(0,\pi)$ has a Fourier sine series that converges to it in $L_2(0,\pi)$; the Fourier sine series of any piecewise continuous function $f(x)$ which has bounded variation on $(0,\pi)$ converges to $\frac{1}{2}[f(x+)+f(x-)]$ throughout $(0,\pi)$ (see § 3).

A spectral approximation is gotten by simply truncating (1.2) to

$$u_N(x, t) = \sum_{n=1}^{N} a_n(t) \sin nx \tag{1.5}$$

and replacing (1.3) by the evolution equation

$$\frac{da_n}{dt} = -n^2 a_n, \qquad n = 1, \cdots, N. \tag{1.6}$$

with the initial conditions $a_n(0) = f_n$ $(n = 1, \cdots, N)$.

The spectral approximation (1.5)–(1.6) to (1.1) is an exceedingly good approximation for any $t > 0$ as $N \to \infty$. In fact, the error $u(x, t) - u_N(x, t)$ goes to zero more rapidly than $e^{-N^2 t}$ as $N \to \infty$ for any $t > 0$. In contrast, a finite difference approximation to the heat equation using N grid points in x but leaving t as a continuous variable (a "semi-discrete" approximation) leads to errors that decay only algebraically with N as $N \to \infty$. (Of course, if we solve (1.6) by finite differences in t the error of the spectral method would go to zero algebraically with the time step Δt. However, we shall neglect all time differencing errors for now and study only the convergence of semi-discrete approximations. Time-differencing methods are discussed in § 9.)

EXAMPLE 1.2. *Fourier sine series solution of an inhomogeneous heat equation.* Not all spectral methods work as well as the trivial one just outlined in Example 1.1. Consider for example the solution to the problem

$$\frac{\partial u}{\partial t} = \frac{\partial^2 u}{\partial x^2} + 1, \qquad 0 < x < \pi, \quad t \geqq 0,$$

with the same initial and boundary conditions as before. The Fourier sine coefficients of the exact solution are now

$$a_n(t) = f_n e^{-n^2 t} + \frac{4}{\pi n^3}(1 - e^{-n^2 t}) e_n \tag{1.7}$$

where $e_n = 0$ if n is even and $e_n = 1$ if n is odd. Spectral approximations are now given by (1.5) with (1.6) replaced by

$$\frac{da_n}{dt} = -n^2 a_n + \frac{4}{\pi n} e_n, \qquad n = 1, \cdots, N,$$

the solution of which is (1.7) for $n = 1, \cdots, N$. Now the truncation error $u(x, t) - u_N(x, t)$ no longer decays exponentially as $N \to \infty$; the error is of order N^{-3} as $N \to \infty$ for fixed x, $0 < x < \pi$, and $t > 0$. In other words, the results to be anticipated from this spectral method behave asymptotically as $N \to \infty$ in the same way as those obtained by a third-order finite-difference scheme (in which the error goes to zero like $\Delta x^3 = (\pi/N)^3$). For this problem, straightforward solution by finite differences may be more efficient and accurate than solution by Fourier series.

The last example may be disturbing but even more serious difficulties confront the unwary user of spectral methods, as the next example should make amply clear.

EXAMPLE 1.3. *Fourier sine series solution of the one-dimensional wave equation.* Consider the mixed initial-boundary value problem for the one-dimensional wave equation

$$\frac{\partial u(x,t)}{\partial t}+\frac{\partial u(x,t)}{\partial x}=x+t, \qquad 0<x<\pi, \quad t\geqq 0, \tag{1.8a}$$

$$u(0,t)=0, \qquad t\geqq 0, \tag{1.8b}$$

$$u(x,0)=0, \qquad 0\leqq x\leqq \pi. \tag{1.8c}$$

The exact solution to this well posed problem is $u(x, t)=xt$. This solution can also be found by Fourier sine series expansion of $u(x, t)$. To do this, we substitute (1.2) into (1.8) and re-expand all terms in sine series. The Fourier expansion of $\partial u/\partial x$ is

$$\frac{\partial u}{\partial x}=\sum_{n=1}^{\infty} b_n(t)\sin nx \tag{1.9}$$

where integration by parts gives

$$\begin{aligned} b_n(t)&=\frac{2}{\pi}\int_0^{\pi}\frac{\partial u}{\partial x}\sin nx\,dx=-\frac{2n}{\pi}\int_0^{\pi}u\cos nx\,dx\\ &=-\frac{2n}{\pi}\sum_{m=1}^{\infty}a_m(t)\int_0^{\pi}\sin mx\cos nx\,dx \qquad (1.10)\\ &=\frac{4}{\pi}\sum_{\substack{m=1\\ m+n \text{ odd}}}^{\infty}\frac{nm}{n^2-m^2}a_m(t). \end{aligned}$$

Also, the Fourier sine coefficients of x are $2/n(-1)^{n+1}$ and the Fourier sine coefficients of t are $(4t/(\pi n))e_n$, where $e_n=0$ if n is even and $e_n=1$ if n is odd. Equating coefficients of $\sin nx$ in (1.8a), we obtain

$$\frac{da_n}{dt}=-\frac{4}{\pi}\sum_{\substack{m=1\\ m+n \text{ odd}}}^{\infty}\frac{nm}{n^2-m^2}a_m-\frac{2}{n}(-1)^n+\frac{4}{\pi n}te_n, \qquad n=1,2,\cdots. \tag{1.11}$$

The Fourier sine coefficients of the exact solution $u(x, t)=xt$ are

$$a_n(t)=-\frac{2}{n}(-1)^n t, \qquad n=1,2,\cdots.$$

It is easy to verify by direct substitution that these coefficients exactly satisfy (1.11); in particular, the sum in (1.11) converges for all t.

Now suppose we employ a spectral method based on Fourier sine series to solve this problem. We seek a solution to (1.8) in the form of the truncated sine series (1.5). If the exact coefficients $a_n(t)$ are used in (1.5) then $u(x, t)-u_N(x, t)\to 0$ as

$N \to \infty$; for each fixed x, $0 < x < \pi$, and $t > 0$ the error is of order $1/N$ as $N \to \infty$ (see § 3).

However, it is not reasonable to assume that the expansion coefficients $a_n(t)$ are known exactly in this case because of the complicated couplings between various n in the system (1.11). It is more reasonable to determine them by numerical solution of an approximation to (1.11). Galerkin approximation (see § 2) gives the truncated system of equations

$$\frac{da_n}{dt} = -\frac{4}{\pi} \sum_{\substack{m=1 \\ m+n \text{ odd}}}^{N} \frac{nm}{n^2 - m^2} a_m - \frac{2}{n}(-1)^n + \frac{4}{\pi n} te_n, \quad n = 1, \cdots, N. \tag{1.12}$$

The truncation of the infinite system (1.11) to the finite system (1.12) is a standard way to approximate infinite coupled systems. Unfortunately, it need not work. In Figs. 1.1–1.2 we show plots of the approximations $u_N(x, t)$ at $t = 5$ given by (1.5) for $N = 50, 75$. These plots are obtained by numerical solution of (1.12) with $a_n(0) = 0$; the time steps used in the numerical solution of (1.12) are

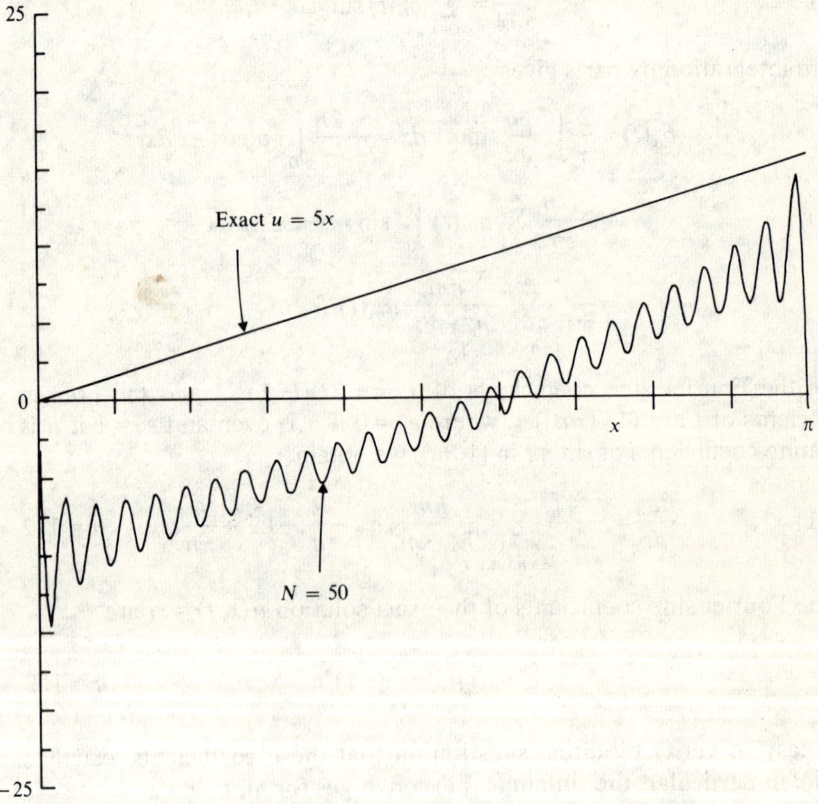

FIG. 1.1. *A plot of the Galerkin approximation* $u_N(x, t)$ *to* (1.8) *for* $N = 50$ *at* $t = 5$. *This solution is obtained by numerical integration of* (1.12). *Time differencing errors are negligible. The exact solution* $u = xt$ *at* $t = 5$ *is also shown.*

so small that time differencing errors are negligible. It is apparent that the approximate solutions with N finite do not converge to the exact solution as N increases! The divergence of this spectral method will be explained in § 6.

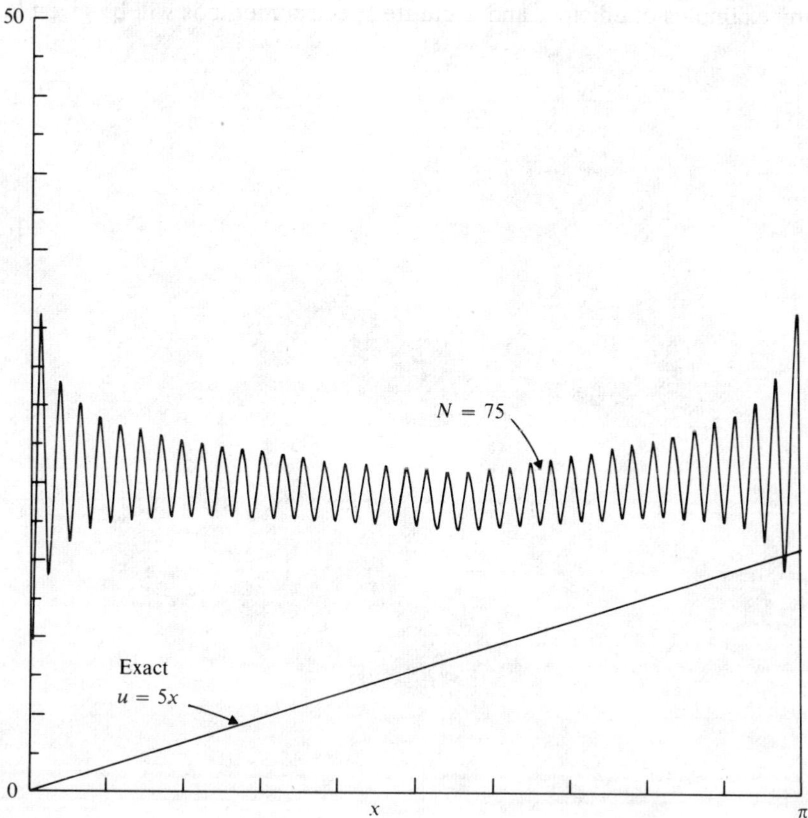

FIG. 1.2. *A plot of the Galerkin approximation* $u_N(x, t)$ *to* (1.8) *for* $N=75$ *at* $t=5$. *This solution is obtained by numerical integration of* (1.12). *Time differencing errors are negligible. The exact solution* $u=xt$ *at* $t=5$ *is also shown.*

Not all spectral methods give such poor results. A properly formulated and implemented spectral method gives results of striking accuracy with efficient use of computer resources. The choice of an appropriate spectral method is governed by two main considerations:

(i) Accuracy. In order to be useful a spectral method should be designed to give results of greater accuracy than can be obtained by more conventional difference methods using similar spatial resolution or degrees of freedom. The choice of appropriate spectral representation depends on the kind of boundary conditions involved in the problem.

(ii) Efficiency. In order to be useful the spectral method should be as efficient as difference methods with comparable numbers of degrees of freedom. For similar work, spectral methods should produce more accurate results than conventional methods.

In § 15, we present a catalog of different spectral methods and indicate the kinds of problems to which they can be most usefully applied.

Many examples of efficient and accurate spectral methods will be given later.

SECTION 2

Spectral Methods

The problems to be studied here are mixed initial-boundary value problems of the form

$$\frac{\partial u(x,t)}{\partial t} = L(x,t)u(x,t) + f(x,t), \qquad x \in D, \quad t \geqq 0, \tag{2.1}$$

$$B(x)u(x,t) = 0, \qquad x \in \partial D, \quad t > 0, \tag{2.2}$$

$$u(x,0) = g(x), \qquad x \in D, \tag{2.3}$$

where D is a spatial domain with boundary ∂D, $L(x,t)$ is a linear (spatial) differential operator and $B(x)$ is a linear (time-independent) boundary operator. Here we write (2.1)–(2.3) for a single dependent variable u and a single space coordinate x with the understanding that much of the following analysis generalizes to systems of equations in higher space dimensions. Also, attention is restricted to problems with homogeneous boundary conditions because the solution to any problem involving inhomogeneous boundary conditions is the sum of an arbitrary function having the imposed boundary values and a solution to a problem of the form (2.1)–(2.3). The extension to nonlinear problems will be indicated in Examples 2.9–2.10 at the end of this section.

Before discussing spectral methods for solution of (2.1)–(2.3) let us set up the mathematical framework for our later analysis. It is assumed that, for each t, $u(x,t)$ is an element of a Hilbert space $\mathcal{H}$ with inner product $(\cdot,\cdot)$ and norm $\|\cdot\|$. For each $t>0$, the solution[1] $u(t)$ belongs to the subspace $\mathcal{B}$ of $\mathcal{H}$ consisting of all functions $u \in \mathcal{H}$ satisfying $Bu = 0$ on ∂D. We do not require that $u(x,0) = g(x) \in \mathcal{B}$ but only that $u(x,0) \in \mathcal{H}$. The operator L is typically an unbounded differential operator whose domain is dense in, but smaller than, $\mathcal{H}$. For example, if $L = \partial/\partial x$ and $\mathcal{H} = L_2(0,1)$, the domain of L can be chosen as the dense set of all absolutely continuous functions on $0 \leqq x \leqq 1$.

If the problem (2.1)–(2.3) is well posed, the evolution operator is a bounded linear operator from $\mathcal{H}$ to $\mathcal{B}$. Boundedness implies that the domain of the evolution operator can be extended in a standard way from the domain of L to the whole space $\mathcal{H}$ (Richtmyer and Morton (1967, p. 34)). For notational convenience we shall assume henceforth that L is time independent so that the evolution operator is $\exp(Lt)$. In this case the formal solution of (2.1)–(2.3) is

$$u(t) = e^{Lt}u(0) + \int_0^t e^{L(t-s)}f(s)\,ds. \tag{2.4}$$

[1] We will often denote $u(x,t)$ by $u(t)$ when discussing u as a function of t.

This formal solution is justified under the conditions that $f(t)$, $Lf(t)$, and $L^2f(t)$ exist and are continuous functions of t in the norm $\|\cdot\|$ for all $t \geqq 0$ (see Richtmyer and Morton (1967)).

The semi-discrete approximations to (2.1) to be studied here are of the form

$$\frac{\partial u_N(x,t)}{\partial t} = L_N u_N(x,t) + f_N(x,t) \tag{2.5}$$

where, for each t, $u_N(x,t)$ belongs to an N-dimensional subspace $\mathcal{B}_N$ of $\mathcal{B}$, and L_N is a linear operator from $\mathcal{H}$ to $\mathcal{B}_N$ of the form

$$L_N = P_N L P_N. \tag{2.6}$$

Here P_N is a projection operator of $\mathcal{H}$ onto $\mathcal{B}_N$ and $f_N = P_N f$. We shall assume that $\mathcal{B}_N \subset \mathcal{B}_M$ when $N < M$. To be definite, we shall also assume the initial conditions for the approximate equations (2.5) to be $u_N(0) = P_N u(0)$, where $u(0) = g(x)$ is the initial condition (2.3). Specific examples of projections P_N and the resulting approximations L_N will be given below.

According to this general framework, the formulation of a spectral method involves two essential steps: (i) the choice of approximation space $\mathcal{B}_N$; and (ii) the choice of the projection operator P_N. It will turn out that the mathematical analysis of the methods also involves two key steps: (i) the analysis of how accurately functions in $\mathcal{H}$ can be approximated by functions in $\mathcal{B}_N$ (see § 3) and, in particular, the estimation of $\|u - P_N u\|$ for arbitrary $u \in \mathcal{H}$; and (ii) the study of the "stability" of L_N (see § 4). Finally, there are the important practical questions of how to discretize time (see § 9) and how to implement spectral methods efficiently (see § 10). All these considerations will be reviewed in § 15.

Galerkin approximation. A Galerkin approximation to (2.1)–(2.3) is constructed as follows. The approximation u_N is sought in the form of the truncated series

$$u_N(x,t) = \sum_{n=1}^{N} a_n(t)\phi_n(x) \tag{2.7}$$

where the time-independent functions ϕ_n are assumed linearly independent and $\phi_n \in \mathcal{B}_N$ for all n. Thus, $u_N(x,t)$ necessarily satisfies all the boundary conditions. The expansion coefficients $a_n(t)$ are determined by the Galerkin equations

$$\frac{d}{dt}(\phi_n, u_N) = (\phi_n, Lu_N) + (\phi_n, f), \qquad n = 1, \cdots, N, \tag{2.8}$$

or

$$\sum_{m=1}^{N} (\phi_n, \phi_m)\frac{da_m}{dt} = \sum_{m=1}^{N} a_m(\phi_n, L\phi_m) + (\phi_n, f).$$

These implicit equations for $a_n(t)$ can be put into the standard explicit form (2.5)–(2.6) by defining the projection operator P_N by

$$P_N u(x) = \sum_{n=1}^{N} \sum_{m=1}^{N} p_{nm}(\phi_m, u)\phi_n(x), \tag{2.9}$$

where p_{nm} are the elements of the inverse of the $N\times N$ matrix whose entries are (ϕ_n, ϕ_m).

Note that the relation

$$P_N u = \sum_{n=1}^{N} \sum_{m=1}^{N} p_{nm}(\phi_m, P_N u)\phi_n(x)$$

holds for all projection operators P_N. However, the specific projection operator (2.9) is particular to Galerkin approximation.

The Galerkin equations (2.8) may be characterized as follows. At each instant t, we assume that the expansion coefficients $a_n(t)$ in (2.7) are known and seek values for the N independent quantities da_n/dt $(n=1,\cdots,N)$ that minimize

$$\left(\frac{\partial u_N}{\partial t} - Lu_N, \frac{\partial u_N}{\partial t} - Lu_N\right).$$

The resulting equations for da_n/dt are (2.8).

EXAMPLE 2.1. *Fourier sine series.* If we choose $\mathscr{H} = L_2(0, \pi)$ and $\phi_n(x) = \sin nx$, we recover the Galerkin approximations given in Examples 1.1-2 for the heat equation and in Example 1.3 for the wave equation. Every function $u \in L_2(0, \pi)$ has a Fourier sine series that converges in the L_2 norm, so that $\|u - P_N u\| \to 0$ as $N \to \infty$. However, as illustrated by Example 1.3, this does not ensure that the Galerkin approximation u_N converges to u as $N \to \infty$.

EXAMPLE 2.2. *Chebyshev series.* We choose $\mathscr{H}$ to be the space of functions on the interval $|x| \leqq 1$ that are square integrable with respect to the weight function $1/\sqrt{1-x^2}$. If the problem is

$$u_t + u_x = f(x, t), \qquad -1 \leqq x \leqq 1, \quad t > 0, \tag{2.10a}$$

$$u(-1, t) = 0, \qquad u(x, 0) = g(x), \tag{2.10b}$$

which is a slight generalization of Example 1.3, it is appropriate to choose the expansion functions for the Galerkin approximates to be $\phi_n(x) = T_n(x) - (-1)^n T_0(x)$. Here $T_n(x)$ is the Chebyshev polynomial of degree n defined by $T_n(\cos\theta) = \cos n\theta$ when $x = \cos\theta$; thus, $T_0(x) = 1$, $T_1(x) = x$, $T_2(x) = 2x^2 - 1$, $T_3(x) = 4x^3 - 3x, \cdots$. Observe that $\phi_n(x)$ satisfies the boundary condition $\phi_n(-1) = 0$ because $T_n(-1) = (-1)^n$ for all n. The properties of Chebyshev polynomials are summarized in the Appendix.

The Galerkin equations (2.8) are obtained explicitly as follows. First, the definition of $T_n(x)$ and the substitution $x = \cos\theta$ imply that

$$(T_n, T_m) = \int_0^{\pi} \cos n\theta \cos m\theta \, d\theta = \frac{\pi}{2} c_n \delta_{nm},$$

where

$$(f, g)=\int_{-1}^{1} f(x)g(x)/\sqrt{1-x^2}\,dx.$$

Here $c_0=2$, $c_n=1$ $(n>0)$ and $\delta_{nm}=0$ if $n\neq m$, 1 if $n=m$. Therefore,

$$(\phi_n, \phi_m)=\frac{\pi}{2}\delta_{nm}+(-1)^{n+m}\pi.$$

Next, the Chebyshev polynomials satisfy

$$2T_n(x)=\frac{T'_{n+1}(x)}{n+1}-\frac{T'_{n-1}(x)}{n-1}, \qquad n\geqq 2,$$

as may be verified by substituting $x=\cos\theta$. Therefore,

$$(\phi_n, \phi'_m)=\begin{cases}\pi(-1)^{n+1}m+\pi m, & n<m, \quad m+n \text{ odd},\\ \pi(-1)^{n+1}m, & n>m, \quad m+n \text{ odd},\\ 0 & n+m \text{ even}.\end{cases}$$

Using these results, (2.8) gives the Galerkin approximation equations

$$\frac{da_n}{dt}+2(-1)^n\frac{d}{dt}\sum_{m=1}^{N}(-1)^m a_m=-2\sum_{\substack{p=n+1\\ p+n \text{ odd}}}^{N} pa_p$$

$$+2(-1)^n\sum_{\substack{p=1\\ p \text{ odd}}}^{N} pa_p+\hat{f}_n+2(-1)^n\hat{f}_0, \qquad n=1,\cdots,N.$$

Here $\hat{f}_n=(T_n, f)$ for $n=0,\cdots,N$.

These Galerkin equations can be simplified by introducing the notation $a_0=-\sum_{m=1}^{N}(-1)^m a_m$, so that (2.7) becomes

$$u_N(x,t)=\sum_{n=0}^{N} a_n(t)T_n(x). \tag{2.11}$$

Substituting the above expression for a_0, the Galerkin equations for a_n can be rewritten as

$$\frac{da_n}{dt}=-\frac{2}{c_n}\sum_{\substack{p=n+1\\ p+n \text{ odd}}}^{N} pa_p+\hat{f}_n+\frac{1}{c_n}b(t)(-1)^n, \qquad n=0,\cdots,N, \tag{2.12}$$

$$\sum_{n=0}^{N}(-1)^n a_n=0. \tag{2.13}$$

Here $b(t)$ is a "boundary" term that ensures maintenance of the boundary condition (2.13). By using (2.13) it is easy to show that the explicit form of $b(t)$ is

$$b(t)=\frac{-1}{N+1/2}\left[\sum_{n=0}^{N}(-1)^n(n^2a_n+\hat{f}_n)\right]=\frac{1}{N+1/2}\left[\frac{\partial u_N}{\partial x}\bigg|_{x=-1}-\sum_{n=0}^{N}(-1)^n\hat{f}_n\right].$$

Tau approximation. The tau method was invented by Lanczos in 1938 (see Lanczos (1956)). First, the expansion functions ϕ_n $(n=1, 2, \cdots)$ are assumed to be elements of a complete set of orthonormal functions. The approximate solution $u_N(x, t)$ is assumed to be expanded in terms of those functions as in

$$u_N(x, t) = \sum_{n=1}^{N+k} a_n(t)\phi_n(x). \tag{2.14}$$

Here k is the number of independent boundary constraints $Bu_N = 0$ that must be applied. The important difference between (2.14) for tau approximation and (2.7) for Galerkin approximation is that the expansion functions ϕ_n in (2.14) are *not* required individually to satisfy the boundary constraints (2.2). The k boundary constraints

$$\sum_{n=1}^{N+k} a_n B\phi_n = 0 \tag{2.15}$$

are imposed as part of the conditions determining the expansion coefficients a_n of a function in $\mathcal{B}_N$. Then, the projection operator P_N is defined by

$$P_N\left(\sum_{n=1}^{\infty} A_n\phi_n\right) = \sum_{n=1}^{N} A_n\phi_n + \sum_{m=1}^{k} b_m\phi_{N+m}, \tag{2.16}$$

where b_m $(m=1, \cdots, k)$ are chosen so that the boundary constraints are satisfied: $BP_N u = 0$ for all $u \in \mathcal{H}$.

It follows from these definitions that the tau approximation to (2.1)–(2.2) is given by (2.14) with the k equations (2.15) and the N equations

$$\frac{da_n}{dt} = (\phi_n, Lu_N) + (\phi_n, f), \qquad n = 1, \cdots, N. \tag{2.17}$$

An equivalent formulation of the tau method is given as follows: The equations for the expansion coefficients a_n of the *exact* solution u in terms of the complete orthonormal basis ϕ_n are

$$u(x, t) = \sum_{n=1}^{\infty} a_n(t)\phi_n(x),$$

$$\frac{da_n}{dt} = (\phi_n, Lu) + (\phi_n, f), \qquad n = 1, 2, \cdots. \tag{2.18}$$

The tau approximation equations for the $N+k$ expansion coefficients of u_N in (2.14) are obtained from the first N equations (2.18) with u replaced by u_N and the k boundary conditions (2.15).

The origin of the name "tau method" is that the resulting approximation u_N is the *exact* solution to the modified problem

$$\frac{\partial u_N}{\partial t} = Lu_N + f + \sum_{p=1}^{\infty} \tau_p(t)\phi_{N+p}(x) \tag{2.19}$$

which lies in $\mathscr{B}_N$ for all $t>0$. For each initial value problem and choice of orthonormal basis ϕ_n (and associated inner product), there is a (normally unique) choice of τ-coefficients such that $u_N \in \mathscr{B}_N$, namely

$$\tau_p = -(\phi_{N+p}, Lu_N + f), \qquad p = k+1, k+2, \cdots.$$

The remaining tau coefficients $\tau_1, \tau_2, \cdots, \tau_k$ are determined by the k boundary constraints

$$B\frac{\partial u_N}{\partial t} = 0$$

and the N dynamical constraints (2.17) for $n = 1, \cdots, N$.

EXAMPLE 2.3. *Fourier sine series.* For all of the applications given in Example 2.1, Galerkin and tau approximations based on $\phi_n = \sqrt{2/\pi}\sin nx$ are identical (except for the scaling factor $\sqrt{2/\pi}$) since the orthonormal expansion functions ϕ_n satisfy the boundary conditions.

EXAMPLE 2.4. *Chebyshev series.* If we choose $\phi_{n+1}(x) = \sqrt{2/(\pi c_n)}\,T_n(x)$ where $c_0 = 2$, $c_n = 1$ $(n>0)$ and apply the tau method to the problem (2.10) the result can be recast into the form of equations (2.11)–(2.13) with $b(t) = 0$ and (2.12) only applied for $n = 0, 1, \cdots, N-1$ instead of $n = 0, 1, \cdots, N$. Thus, the tau equations for the one-dimensional wave problem (2.10) are (2.11) with

$$\frac{da_n}{dt} = -\frac{2}{c_n}\sum_{\substack{p=n+1\\ p+n\ \text{odd}}}^{N} pa_p + \hat{f}_n, \qquad 0 \leqq n \leqq N-1, \tag{2.20}$$

$$\sum_{n=0}^{N} (-1)^n a_n(t) = 0. \tag{2.21}$$

In this problem, $\sqrt{2/\pi}\,\tau_1(t) = a_N' - \hat{f}_N$ while $\tau_p(t) \equiv 0$ for $p>1$.

EXAMPLE 2.5. *Laguerre series.* Here we choose $\mathscr{H}$ to be the space of functions that are square integrable on $0 \leqq x < \infty$ with respect to the weight function e^{-x}. We choose the expansion functions to be $\phi_n(x) = L_n(x)$ where $L_n(x)$ is the (normalized) Laguerre polynomial of degree n: $L_0(x) = 1$, $L_1(x) = 1-x$, $L_2(x) = 1 - 2x + \frac{1}{2}x^2, \cdots$.

Suppose we wish to solve the problem

$$u_t + u_x = f(x, t), \qquad 0 \leqq x < \infty, \quad t>0, \tag{2.22a}$$

$$u(0, t) = 0, \qquad u(x, 0) = g(x) \tag{2.22b}$$

by seeking an approximate solution of the form

$$u_N(x, t) = \sum_{n=0}^{N} a_n(t) L_n(x). \tag{2.23}$$

To derive the tau equations for $a_n(t)$, we note that $L_n(x)$ satisfies $L_n(0) = 1$, $L_n' - L_{n+1}' = L_n$, $n = 0, 1, \cdots$ and $(L_n, L_m) \equiv \int_0^\infty L_n(x) L_m(x) e^{-x}\,dx = \delta_{nm}$. Thus, the tau approximation (2.17) is

$$\frac{da_n}{dt}=\sum_{p=n+1}^{N} a_p+(L_n, f), \qquad n=0, 1, \cdots, N-1, \tag{2.24}$$

while the boundary condition is

$$\sum_{n=0}^{N} a_n=0. \tag{2.25}$$

Similarly, the Laguerre-tau approximation to the heat equation problem

$$\begin{aligned} &u_t=u_{xx}+f(x, t), \qquad 0\leqq x<\infty, \quad t>0,\\ &u(0, t)=0, \qquad u(x, 0)=g(x) \end{aligned} \tag{2.26}$$

is given by (2.23), (2.25) and

$$\frac{da_n}{dt}=\sum_{p=n+1}^{N} (p-n-1)a_p+(L_n, f), \qquad n=0, 1, \cdots, N-1. \tag{2.27}$$

Collocation or pseudospectral approximation. The projection operator P_N for collocation [sometimes called the method of selected points (Lanczos (1956)) or pseudospectral approximation (Orszag (1971c))] is defined as follows. Let $x_1, x_2, \cdots, x_N$ be N points interior to the domain D. These points are called the collocation points. Also let $\phi_n(x)$ $(n=1, \cdots, N)$ be a basis for the approximation space $\mathscr{B}_N$ and suppose that $\det \phi_n(x_m)\neq 0$. Then for each $u\in \mathscr{H}$,

$$P_N u=\sum_{n=1}^{N} a_n\phi_n(x), \tag{2.28}$$

where the expansion coefficients a_n are the solutions of the equations

$$\sum_{n=1}^{N} a_n\phi_n(x_i)=u(x_i), \qquad i=1, \cdots, N. \tag{2.29}$$

Thus, collocation is characterized by the conditions that $P_N u(x_i)=u(x_i)$ for $i=1, \cdots, N$ and $P_N u\in \mathscr{B}_N$. Notice that the results of collocation depend on both the points x_n and the functions $\phi_n(x)$ for $n=1, \cdots, N$.

EXAMPLE 2.6. *Fourier sine series.* If we wish to solve the problems formulated in Examples 1.1–1.3 by collocation instead of Galerkin or tau methods we proceed as follows. We choose the space $\mathscr{H}=L_2(0, \pi)$, the expansion functions $\phi_n(x)=\sin nx$ $(n=1, \cdots, N)$, and the collocation points $x_j=\pi j/(N+1)$, $(j=1, \cdots, N)$. The collocation equations

$$\sum_{n=1}^{N} a_n \sin\frac{\pi jn}{N+1}=u(x_j), \qquad j=1, \cdots, N \tag{2.30}$$

have the explicit solution

$$a_n=\frac{2}{N+1}\sum_{j=1}^{N} u(x_j)\sin\frac{\pi jn}{N+1}, \qquad n=1, \cdots, N. \tag{2.31}$$

This result follows from the relation

$$\sum_{n=1}^{N} \sin\frac{\pi jn}{N+1}\sin\frac{\pi kn}{N+1}=\frac{N+1}{2}\delta_{jk}$$

valid for $0<j,\ k<N+1$. Thus,

$$P_N u = \sum_{n=1}^{N} a_n \sin nx \tag{2.32}$$

where a_n is given by (2.31).

It follows from (2.30)–(2.32) that

$$P_N L P_N u = \sum_{n=1}^{N} b_n \sin nx$$

where $b_n = -n^2 a_n$ $(n = 1, \cdots, N)$ if $L = \partial^2/\partial x^2$, and

$$b_n = \frac{2}{N+1} \sum_{\substack{m=1\\ m+n \text{ odd}}}^{N} \frac{m \sin[\pi n/(N+1)]}{\cos[\pi m/(N+1)] - \cos[\pi n/(N+1)]} a_m, \qquad n = 1, \cdots, N,$$

if $L = \partial/\partial x$. [Note that, as $N\to\infty$, the latter expression reproduces (1.10).]

EXAMPLE 2.7. *Chebyshev collocation for the wave equation.* Suppose we wish to solve the one-dimensional wave problem (2.10) using collocation. An appropriate basis for the approximation space $\mathcal{B}_N$ is the set of functions $\phi_n(x) = T_n(x) - (-1)^n T_0(x)$ $(n = 1, \cdots, N)$ introduced in our discussion of Example 2.2 above. We choose the collocation points to be the extrema of the Chebyshev polynomial $T_N(x)$ satisfying $|x| \leqq 1$. Since $T_N(\cos\theta) = \cos N\theta$, these extrema lie at $x_j = \cos(\pi j/N)$ for $j = 0, \cdots, N-1$. The point $x_N = -1$ is also an extremum of $T_N(x)$ but it is *not* included in the set of collocation points because the boundary conditions for (2.10) are imposed at $x = -1$ so $\phi_n(-1) = 0$ for all n.

As in Example 2.2, the expansion coefficients a_n for $n = 1, \cdots, N$ may be augmented by defining $a_0 = -\sum_{m=1}^{N} (-1)^m a_m$ so that

$$u_N(x, t) = \sum_{n=0}^{N} a_n(t) T_n(x).$$

It may then be shown that the collocation equations for $a_n(t)$ that follow from (2.10) are

$$\frac{da_n}{dt} = -\frac{2}{c_n} \sum_{\substack{p=n+1\\ p+n \text{ odd}}}^{N} p a_p + \hat{f}_n + \frac{1}{\bar{c}_n} b(t)(-1)^n, \qquad n = 0, \cdots, N, \tag{2.33}$$

$$\sum_{n=0}^{N} (-1)^n a_n(t) = 0, \tag{2.34}$$

where $\hat{f}_n = (T_n, f)$ and $\bar{c}_0 = \bar{c}_N = 2$, $\bar{c}_n = 1$ $(0<n<N)$. Here $b(t)$ is a "boundary" term that is used to ensure compliance with the boundary condition (2.34). It may

also be shown that

$$b(t)=-\frac{1}{N}\sum_{n=0}^{N}(-1)^n(n^2a_n+\hat{f}_n)=\frac{1}{N}\left[\frac{\partial u_N}{\partial x}\bigg|_{x=-1}-\sum_{n=0}^{N}(-1)^n\hat{f}_n\right].$$

The reader should observe the close similarity between the Chebyshev Galerkin, tau, and collocation equations for the problem (2.10). The only difference between them is the way the boundary term $b(t)$ enters. In the Galerkin equations (2.12), $b(t)$ appears with the coefficient $(-1)^n/c_n$; in the tau equations $b(t)$ enters with the coefficient δ_{nN} so it appears only in the equation for a_N as a tau coefficient; with collocation, the coefficient of $b(t)$ is $(-1)^n/\bar{c}_n$. This close similarity between the three methods for the wave equation can also be seen by observing that when $f(x, t)$ is a polynomial of degree N in x, all three approximation methods give Nth degree polynomial approximations $u_N(x, t)$ that satisfy *exactly* the initial-boundary value problem

$$\text{(2.35)}\qquad \frac{\partial u_N}{\partial t}+\frac{\partial u_N}{\partial x}=f(x, t)+\tau(t)Q_N(x),$$

$$u_N(-1, t)=0.$$

The tau approximation is obtained if $Q_N(x)=T_N(x)$; the collocation approximation is obtained if

$$Q_N(x)=\prod_{j=0}^{N-1}(x-x_j)=2^{2-N}\sum_{n=0}^{N}\frac{(-1)^{n+N}}{\bar{c}_n}T_n(x)=\frac{1}{N}2^{1-N}(x-1)T_N'(x),$$

where $x_j=\cos(\pi j/N)$ $(j=0,\cdots,N-1)$ are the collocation points; finally, the Galerkin equations (2.12) are obtained if

$$Q_N(x)=\sum_{n=0}^{N}\frac{(-1)^n}{c_n}T_n(x).$$

For all three methods $\tau(t)$ is uniquely determined by the requirement that $u_N(x, t)$ be a polynomial of degree N in x that satisfies the boundary condition $u_N(-1, t)=0$ for all t.

EXAMPLE 2.8. *Chebyshev spectral methods for the heat equation.* To illustrate further the nature of the differences between Galerkin, tau and collocation methods, we apply them to the heat equation

$$\frac{\partial u}{\partial t}=\frac{\partial^2 u}{\partial x^2}+f(x, t),\qquad -1<x<1,\quad t>0,$$

$$u(-1, t)=u(1, t)=0\quad (t>0),\qquad u(x, 0)=g(x)\quad (-1<x<1).$$

We approximate $u(x, t)$ by

$$u_N(x, t)=\sum_{n=0}^{N}a_n(t)T_n(x).$$

The Galerkin, tau, and collocation equations for $a_n(t)$ are all of the form

$$\frac{da_n}{dt}=\frac{1}{c_n}\sum_{\substack{p=n+2\\p+n\text{ even}}}^{N} p(p^2-n^2)a_p+\hat{f}_n(t)+b_1(t)B_{1n}+b_2(t)B_{2n}, \tag{2.36}$$

$$\sum_{n=0}^{N} a_n=\sum_{n=0}^{N}(-1)^n a_n=0, \tag{2.37}$$

where $\hat{f}_n=(T_n, f)$. Equations (2.37) are just a restatement of $u_N(\pm 1, t)=0$. The terms $b_1(t)$ and $b_2(t)$ in (2.36) are boundary terms that ensure compliance with the boundary conditions (2.37). The only differences between the three approximation methods lie in the coefficients B_{1n} and B_{2n}.

In the tau method,

$$B_{1n}=\delta_{n,N-1}, \qquad B_{2n}=\delta_{nN}. \tag{2.38a}$$

In the Galerkin method,

$$B_{1n}=\frac{1}{c_n}, \qquad B_{2n}=\frac{(-1)^n}{c_n}; \tag{2.38b}$$

this result follows using the expansion functions

$$\phi_n(x)=T_n(x)-\begin{cases}T_0(x), & n \text{ even},\\ T_1(x), & n \text{ odd},\end{cases}$$

that satisfy $\phi_n(\pm 1)=0$ and augmenting the expansion coefficients a_n for $n \geqq 2$ by $a_0=-\sum a_{2n}$ and $a_1=-\sum a_{2n+1}$. Finally, with collocation performed at the points $x_j=\cos(\pi j/N)$ $(j=1, 2, \cdots, N-1)$ the coefficients B_{1n} and B_{2n} in (2.36) are given by

$$B_{1n}=\frac{1}{\bar{c}_n}, \qquad B_{2n}=\frac{(-1)^n}{\bar{c}_n}. \tag{2.38c}$$

It may also be verified that the boundary terms $b_1(t)$ and $b_2(t)$ are of the form

$$b_i(t)=c_{i+}\left[\left.\frac{\partial^2 u}{\partial x^2}\right|_{x=+1}+\sum_{n=0}^{N}\hat{f}_n\right]+c_{i-}\left[\left.\frac{\partial^2 u}{\partial x^2}\right|_{x=-1}+\sum_{n=0}^{N}(-1)^n\hat{f}_n\right] \tag{2.39}$$

for $i=1, 2$. Here

$$c_{1+}=-\tfrac{1}{2}, \qquad c_{1-}=\tfrac{1}{2}(-1)^N,$$

$$c_{2+}=-\tfrac{1}{2}, \qquad c_{2-}=\tfrac{1}{2}(-1)^{N+1},$$

for the tau method;

$$c_{1+}=-\frac{N+\frac{1}{2}}{N^2+N}, \qquad c_{1-}=\frac{1}{2}\frac{(-1)^N}{N^2+N},$$

$$c_{2+}=\frac{1}{2}\frac{(-1)^N}{N^2+N}, \qquad c_{2-}=-\frac{N+\frac{1}{2}}{N^2+N},$$

for the Galerkin method;

$$c_{1+} = -\frac{1}{N}, \qquad c_{1-} = 0,$$

$$c_{2+} = 0, \qquad c_{2-} = -\frac{1}{N},$$

for the collocation method.

In the previous examples the only difference between Galerkin, tau, and collocation approximations is their treatment of the boundary terms. However, in more complicated problems, there are significant differences between these approximations. The next example illustrates the influence of quadratic nonlinearity.

EXAMPLE 2.9. *Chebyshev approximations to Burgers' equation.* Chebyshev series approximations to the solution $u(x, t)$ to Burgers' equation

$$\frac{\partial u}{\partial t} + u\frac{\partial u}{\partial x} = \nu \frac{\partial^2 u}{\partial x^2}, \qquad |x| \leqq 1, \quad t > 0, \tag{2.40}$$
$$u(\pm 1, t) = 0, \qquad u(x, 0) = f(x)$$

are obtained by methods very similar to those for linear equations. In general, spectral approximations to the nonlinear equation

$$\frac{\partial u}{\partial t} = A(u) \tag{2.41}$$

are of the form

$$\frac{\partial u_N}{\partial t} = P_N A(P_N u_N), \tag{2.42}$$

where P_N is a projection operator. The projection operator P_N can be that for Galerkin, tau, or collocation approximations.[2]

If we write

$$u_N(x, t) = \sum_{n=0}^{N} a_n(t) T_n(x),$$

---

[2] Observe that if $(u, Au) = 0$ so the system (2.41) has the energy integral $\partial(u, u)/\partial t = 0$, then (2.42) has the energy integral $\partial(u_N, u_N)/\partial t = 0$ provided that the projection operator P_N is self-adjoint. This follows from

$$(u_N, P_N A(P_N u_N)) = (P_N u_N, A(P_N u_N)) = 0.$$

An example of a self-adjoint projection operator P_N is the Galerkin operator (2.9). Energy conservation is guaranteed only if the inner product used in the definition of the Galerkin approximation is the same as that in the energy integral.

then the Galerkin approximation to (2.40) is given by

$$c_n \frac{da_n}{dt} = -2 \sum_{\substack{|m| \leqq N \\ |p| \leqq N \\ m+p \geqq n+1 \\ n+m+p \text{ odd}}} p\bar{a}_m \bar{a}_p + \nu \sum_{\substack{m=n+2 \\ m+n \text{ even}}} m(m^2-n^2) a_m \tag{2.43a}$$

$$+ b_+(t) + b_-(t)(-1)^n, \qquad 0 \leqq n \leqq N,$$

$$\sum_{n=0}^{N} a_n = \sum_{n=0}^{N} a_n(-1)^n = 0, \tag{2.43b}$$

where $\bar{a}_m = c_{|m|} a_{|m|}$ for $|m| \leqq N$. The tau equations are identical except that (2.43a) only applies for $0 \leqq n \leqq N-2$ and $b_+ = b_- = 0$. On the other hand, the collocation equations obtained using the collocation points $x_j = \cos(\pi j/N)$ for $j = 1, \cdots, N-1$ are just (2.43b) and

$$\bar{c}_n \frac{da_n}{dt} = -2 \sum_{\substack{|m| \leqq N \\ |p| \leqq N \\ m+p \geqq n+1 \\ n+m+p \text{ odd}}} p\bar{a}_m \bar{a}_p - 2 \sum_{\substack{|m| \leqq N \\ |p| \leqq N \\ m+p \geqq 2N-n+1 \\ n+m+p \text{ odd}}} p\bar{a}_m \bar{a}_p \tag{2.44}$$

$$+ \nu \sum_{\substack{m=n+2 \\ m+n \text{ even}}}^{N} m(m^2-n^2) a_m + \bar{b}_+(t) + \bar{b}_-(t)(-1)^n, \qquad 0 \leqq n \leqq N,$$

where $\bar{c}_0 = \bar{c}_N = 2$ and $\bar{c}_n = 1$ for $n \neq 0, N$. Observe the appearance of the "aliasing" term as the second sum on the right side of (2.44). Aliasing is discussed in detail by Orszag (1971a), (1972).

EXAMPLE 2.10. *Chebyshev approximations to* $u_t + F(u)_x = 0$. Galerkin and tau approximations to the solution to

$$u_t + F(u)_x = 0 \tag{2.45}$$

where $F(u)$ is arbitrarily nonlinear, are very unwieldy both to write down explicitly and to solve on a computer. On the other hand, while the collocation equations may also be hard to write down explicitly, they lend themselves to ready solution without their explicit form being known!

The collocation approximation to (2.45) is obtained as follows. We use the relation

$$(F(u_N))_x = F'(u_N) \frac{\partial u_N}{\partial x}. \tag{2.46}$$

Since $\partial u_N / \partial x$ can be computed explicitly in terms of u_N as a polynomial in x of degree $N-1$, it follows that $(F(u_N))_x$ can be evaluated by this formula at each of the collocation points assuming that $F'(z)$ is a known function; thus, the collocation approximation to (2.45) is determined.

There is a slightly different collocation procedure that can also be applied to (2.45). It has the operator form

$$\frac{\partial u_N}{\partial t}+P_N\frac{\partial}{\partial x}P_NF(u_N)=0, \tag{2.47}$$

which is usually not the same as the collocation approximation of the form (2.42) described above. In the approximation (2.47), $\partial u_N/\partial t$ is computed by first using collocation to obtain $P_NF(u_N)$ from u_N and then using the collocation approximation $P_N\partial/\partial x$ to $\partial/\partial x$ given in Example 2.7. The collocation approximation given by (2.42) or (2.46) differs from (2.47) by the term

$$P_N\frac{\partial}{\partial x}(I-P_N)F(u_N)$$

which is generally not zero. However, if $F'(z)$ is not known accurately then (2.47) may be the only viable method.

SECTION 3

Survey of Approximation Theory

The remarkable convergence properties of spectral methods to be discussed later owe to the rapid convergence of expansions of smooth functions in series of orthogonal functions. We present a summary of the relevant theory here.

Fourier series. The complex Fourier series of $f(x)$ defined for $0 \leqq x \leqq 2\pi$ is the periodic function

$$g(x) = \sum_{k=-\infty}^{\infty} a_k e^{ikx}, \tag{3.1}$$

where

$$a_k = \frac{1}{2\pi} \int_0^{2\pi} f(x) e^{-ikx} \, dx. \tag{3.2}$$

We shall show below that if $f(x)$ is piecewise continuous and has bounded total variation then

$$g(x) = \tfrac{1}{2}[f(x+) + f(x-)] \tag{3.3}$$

for $0 \leqq x \leqq 2\pi$ and $g(x)$ is repeated periodically outside the interval $0 \leqq x < 2\pi$. In particular, $g(0) = g(2\pi) = \frac{1}{2}[f(0+) + f(2\pi-)]$.

The Fourier sine series of a function $f(x)$ defined for $0 < x < \pi$ is the function

$$g_s(x) = \sum_{k=1}^{\infty} a_k \sin kx, \tag{3.4}$$

where

$$a_k - \frac{2}{\pi} \int_0^{\pi} f(x) \sin kx \, dx. \tag{3.5}$$

The Fourier cosine series of a function defined for $0 < x < \pi$ is

$$g_c(x) = \sum_{k=0}^{\infty} a_k \cos kx, \tag{3.6}$$

where

$$a_k = \frac{2}{\pi c_k} \int_0^{\pi} f(x) \cos kx \, dx \tag{3.7}$$

with $c_0=2$, $c_k=1$ $(k>0)$. It follows easily from (3.3) that if $f(x)$ is piecewise continuous and has bounded total variation then

$$g_s(x)=f_s(x), \tag{3.8}$$

$$g_c(x)=f_c(x), \tag{3.9}$$

where

$$\begin{aligned} &f_s(x)=f_c(x)=\tfrac{1}{2}[f(x+)+f(x-)] \quad \text{for } 0<x<\pi,\\ &f_s(-x)=-f_s(x),\quad f_c(-x)=f_c(x) \quad \textit{for } -\pi<x<0,\\ &f_s(0)=f_s(\pi)=0,\quad f_c(0)=f(0+),\quad f_c(\pi)=f(\pi-). \end{aligned} \tag{3.10}$$

and $f_s(x)$ and $f_c(x)$ are extended periodically outside the interval $-\pi<x\leqq\pi$.

Convergence of Fourier series. To prove (3.3) we define $g_K(x)$ as the partial sum

$$g_K(x)=\sum_{k=-K}^{K} a_k e^{ikx}. \tag{3.11}$$

Using (3.2) and the trigonometric sum formula

$$\sum_{k=-K}^{K} e^{iks}=\frac{\sin[(K+\frac{1}{2})s]}{\sin(\frac{1}{2}s)},$$

we obtain

$$g_K(x)=\frac{1}{2\pi}\int_{x-2\pi}^{x}\frac{\sin[(K+\frac{1}{2})t]}{\sin[\frac{1}{2}t]}f(x-t)\,dt. \tag{3.12}$$

The kernel $\sin(K+\frac{1}{2})t/\sin\frac{1}{2}t$ of the integral (3.12) is plotted for several values of K in Fig. 3.1. This plot suggests that when $f(x)$ has bounded total variation the leading contribution to the integral as $K\to\infty$ comes from the neighborhood of $t=0$ since the contributions from the rest of the integration region should nearly cancel due to the rapid oscillations of the integrand. Thus,

$$g_K(x)\sim\frac{1}{2\pi}\int_{-\varepsilon}^{+\varepsilon}\frac{\sin[(K+\frac{1}{2})t]}{\sin[\frac{1}{2}t]}f(x-t)\,dt, \qquad K\to\infty, \tag{3.13}$$

for any fixed $\varepsilon>0$. Since ε may be chosen small we may replace $\sin\frac{1}{2}t$ by $\frac{1}{2}t$ with a maximum error of $O(\varepsilon^3)$. Also since $f(x-t)$ is piecewise continuous, we may assume that $f(x-t)$ is continuous for $0<t\leqq\varepsilon$ and $-\varepsilon\leqq t<0$ with at worst a jump discontinuity at $t=0$. Therefore we may replace $f(x-t)$ by $f(x-)$ for $t>0$ and $f(x+)$ for $t<0$ giving

$$g_K(x)\sim[f(x+)+f(x-)]\frac{1}{\pi}\int_0^{\varepsilon}\frac{\sin(K+\frac{1}{2})s}{s}\,ds, \qquad K\to\infty.$$

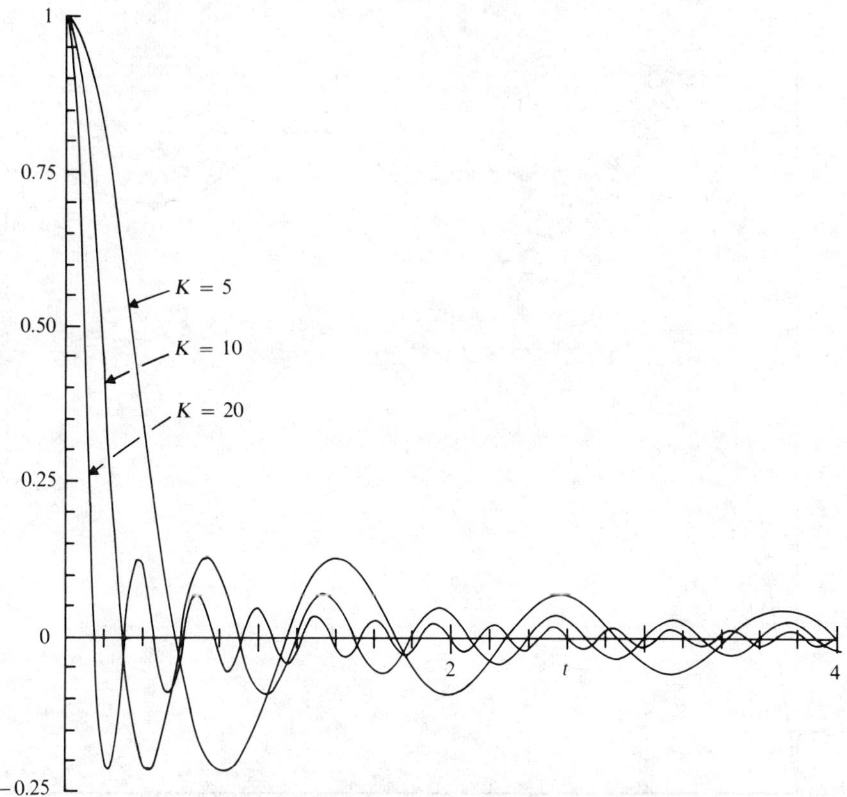

FIG. 3.1. *A plot of* $\sin[(K+\frac{1}{2})t]/[(K+\frac{1}{2})t]$ *for* $K = 5, 10, 20$.

Since

$$\frac{1}{\pi}\int_0^{\varepsilon}\frac{\sin(K+\frac{1}{2})s}{s}\,ds=\frac{1}{\pi}\int_0^{(K+1/2)\varepsilon}\frac{\sin s}{s}\,ds\sim\frac{1}{\pi}\int_0^{\infty}\frac{\sin s}{s}\,ds=\tfrac{1}{2},\qquad K\to\infty,$$

for any fixed $\varepsilon>0$, we obtain

$$g_K(x)\sim\tfrac{1}{2}[f(x+)+f(x-)],\qquad K\to\infty,$$

proving (3.3).

In the neighborhood of a point of discontinuity of $f(x)$ [or $x=0$ and $x=2\pi$ if $f(0+)\neq f(2\pi-)$] the convergence of $g_K(x)$ to its limit (3.3) as $K\to\infty$ is not uniform. To investigate the detailed approach of $g_K(x)$ to $g(x)$ near a point of discontinuity x_0 of $f(x)$, we use the asymptotic integral representation (3.13) to obtain

$$g_K\left(x_0+\frac{z}{K+\frac{1}{2}}\right)\sim\frac{1}{\pi}\int_{-\varepsilon}^{\varepsilon}\frac{\sin[(K+\frac{1}{2})t]}{t}f\left(x_0+\frac{z}{K+\frac{1}{2}}-t\right)dt,\qquad K\to\infty,$$

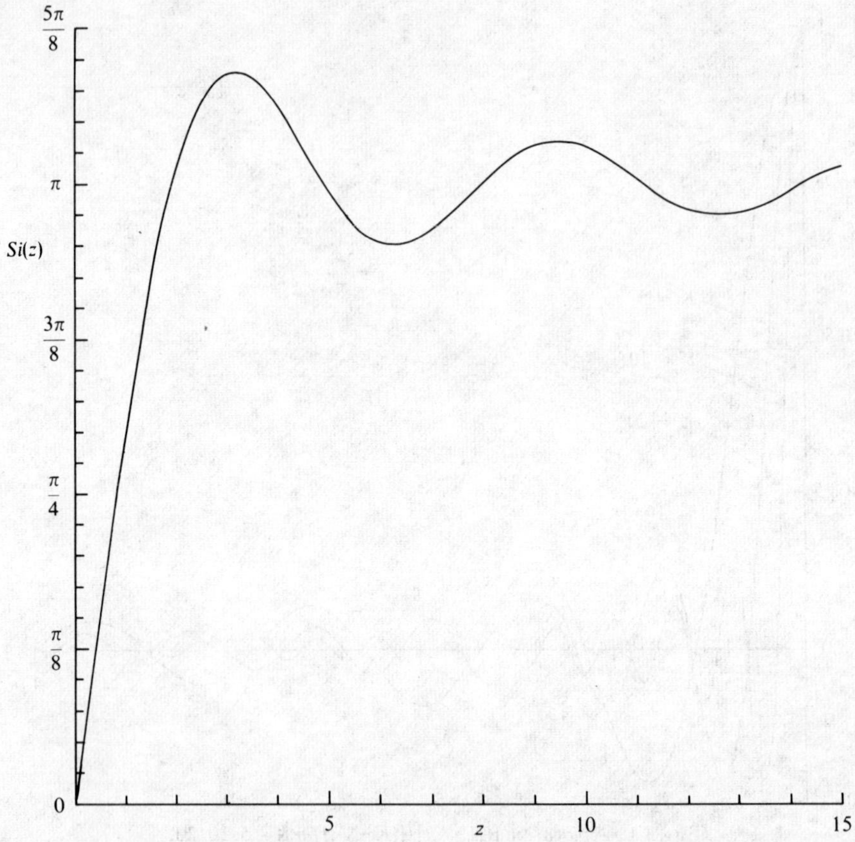

FIG. 3.2. *A plot of the sine integral* Si (z) *defined in* (3.14b) *for* $0 \leqq z \leqq 15$.

for every fixed z. Since ε is assumed small we can approximate $f(x_0+s)$ by $f(x_0+)$ for $0<s<\varepsilon$ and by $f(x_0-)$ for $-\varepsilon<s<0$. Therefore, for each fixed z and ε,

$$g_K\left(x_0+\frac{z}{K+\frac{1}{2}}\right) \sim \frac{f(x_0+)}{\pi}\int_{-\varepsilon}^{z/(K+1/2)} \frac{\sin\,(K+\frac{1}{2})t}{t}\,dt + \frac{f(x_0-)}{\pi}\int_{z/(K+1/2)}^{\varepsilon} \frac{\sin\,(K+\frac{1}{2})t}{t}\,dt \qquad K\to\infty,$$

$$= \frac{f(x_0+)}{\pi}\int_{-\varepsilon(K+1/2)}^{z} \frac{\sin s}{s}\,ds + \frac{f(x_0-)}{\pi}\int_{z}^{\varepsilon(K+1/2)} \frac{\sin s}{s}\,dx, \qquad K\to\infty,$$

$$\sim \frac{f(x_0+)}{\pi}\int_{-\infty}^{z} \frac{\sin s}{s}\,dx + \frac{f(x_0-)}{\pi}\int_{z}^{\infty} \frac{\sin s}{s}\,ds, \qquad K\to\infty.$$

Since $\int_{-\infty}^{\infty} \sin s/s\,ds = \pi$, we obtain

$$\text{(3.14a)} \qquad g_K\left(x_0+\frac{z}{K+\frac{1}{2}}\right) \sim \tfrac{1}{2}[f(x_0+)+f(x_0-)] + \frac{1}{\pi}[f(x_0+)-f(x_0-)]\ \text{Si}\,(z), \qquad K\to\infty,$$

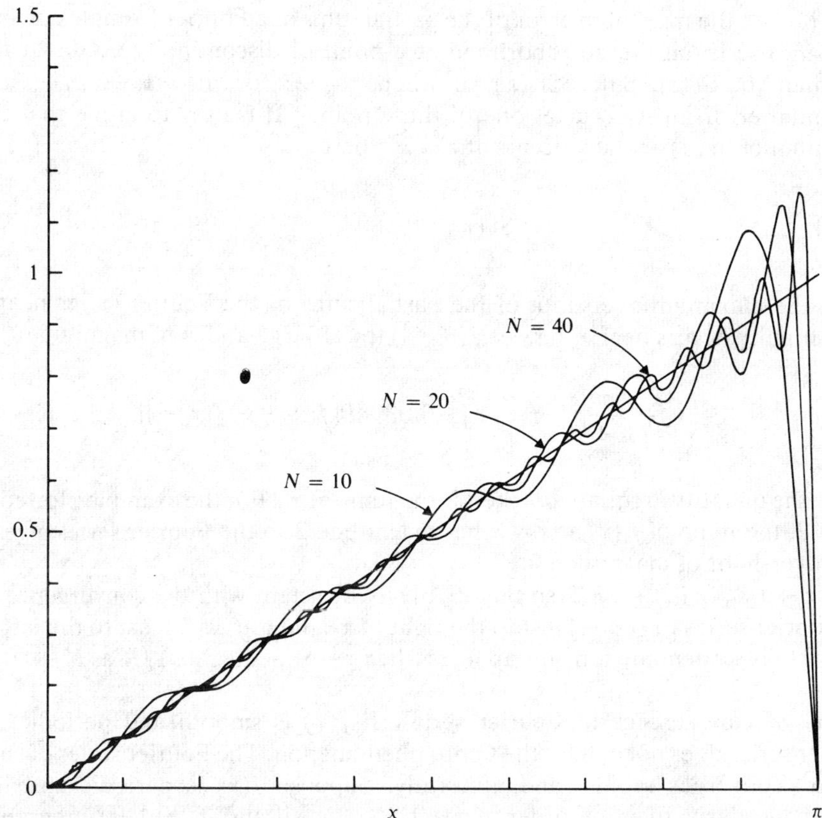

FIG. 3.3. *A plot of the N-term partial sums of the Fourier sine series expansion of the function x/π for $N = 10, 20, 40$. The function x/π is also plotted. Observe the Gibbs phenomenon near $x = \pi$. Also observe that the rate of convergence of the Fourier series is like $1/N$ for x in the interior of the interval $0 < x < \pi$.*

for any fixed z. Here the sine integral Si (z) is defined

$$\text{Si}(z) = \int_0^z \frac{\sin s}{s}\, ds. \tag{3.14b}$$

A plot of Si (z) is given in Fig. 3.2.

The result (3.14) shows that, if $x - x_0 = O(1/K)$ as $K \to \infty$, then $g_K(x)\ \frac{1}{2}[f(x_0+) + f(x_0-)] - O(1)$. This shows the nonuniformity of convergence of $g_K(x)$ to $f(x)$ in the neighborhood of the discontinuity x_0. This nonuniform behavior of the limit $g_K(x) \to f(x)$ as $K \to \infty$ is called the *Gibbs phenomenon*.

To illustrate the Gibbs phenomenon in an actual Fourier series, we plot in Fig. 3.3 the partial sums of the Fourier sine series expansion of the function

$$f(x) = x/\pi, \qquad 0 < x < \pi.$$

The extended function $f_s(x)$ is discontinuous at $x = \pi$ leading to the Gibbs phenomenon there.

As $K \to \infty$, the maximum error of the partial sums of a Fourier (complex or sine or cosine) series in the neighborhood of a point of discontinuity occurs at the maximum of Si (z). Since Si$'(z)=0$ when $z=n\pi$ for $n=\pm 1, \pm 2, \cdots$, the maximum error must occur at one of these points. It is easy to argue that the maximum of Si (z) actually occurs at $z=\pi$ where

$$\frac{1}{\pi}\,\mathrm{Si}\,(\pi) \doteq 0.58949. \tag{3.15}$$

Thus, the maximum overshoot of the partial sums of the Fourier series near a discontinuity occurs near $x = x_0 + \pi/(K+\frac{1}{2})$ for K large and is of magnitude

$$g_K\left(x_0 + \frac{\pi}{K+\frac{1}{2}}\right) - f(x_0+) \sim 0.08949[f(x_0+) - f(x_0-)], \qquad K \to \infty, \tag{3.16}$$

where the quantity in square brackets is the jump at x_0. For the example plotted in Fig. 3.3, the jump of $f_s(x)$ at $x=\pi$ has magnitude 2 so the Fourier series gives a local overshoot of magnitude 0.179.

As $z \to \pm\infty$, Si $(z) \to \pm\pi/2$ so that (3.14) is consistent with the convergence of the Fourier series to $f(x_0+)$ just to the right of x_0 and to $f(x_0-)$ just to the left of x_0. The Gibbs phenomenon only appears when $x \to x_0$ at the rate $1/K$ as $K \to \infty$.

Rate of convergence of Fourier series. If $f(x)$ is smooth and periodic, its Fourier series does not exhibit the Gibbs phenomenon. The Fourier series of such an $f(x)$ converges rapidly and uniformly. Suppose $f(x)$ is periodic and has continuous derivatives of order $p=0, 1, \cdots, n-1$ and $f^{(n)}(x)$ is integrable. Applying integration by parts to (3.2), it follows that

$$a_k = \frac{1}{2\pi(ik)^n}\int_0^{2\pi} f^{(n)}(x)\,e^{-ikx}\,dx.$$

Since $f^{(n)}(x)$ is integrable, the Riemann–Lebesgue lemma implies that

$$a_k \ll 1/k^n, \qquad k \to \pm\infty. \tag{3.17}$$

Note that, because $f(x)$ is periodic, continuity of $f^{(p)}(x)$ also requires $f^{(p)}(0) = f^{(p)}(2\pi)$. It follows from (3.17) that if $f(x)$ is continuous with $f(0)=f(2\pi)$ and $f'(x)$ is integrable then $a_k \ll 1/k$ as $k \to \infty$; if, in addition, $f'(x)$ is *piecewise* continuous and differentiable then $a_k = O(1/k^2)$ as $k \to \infty$.

Now we can be more precise in our estimates of the error $g_K(x) - f(x)$. If a_k goes to zero like $1/k^n$ as $k \to \infty$ and no faster, then $f^{(n-1)}(x)$ is discontinuous. In this case,

$$g_K(x) - f(x) = O\left(\frac{1}{K^n}\right), \qquad K \to \infty, \tag{3.18}$$

when x is fixed away from a point of discontinuity of $f^{(n-1)}$ as $K \to \infty$, while

(3.19) $$g_K(x) - f(x) = O\left(\frac{1}{K^{n-1}}\right), \qquad K \to \infty,$$

when $x - x_0 = O(1/K)$ as $K \to \infty$ where x_0 is a point of discontinuity of $f^{(n-1)}(x)$.

In particular, if $f(x)$ is infinitely differentiable and periodic $[f(x+2\pi) = f(x)]$, $g_K(x)$ converges to $f(x)$ more rapidly than any finite power of $1/K$ as $K \to \infty$ for all x.

Fourier sine and cosine series have convergence properties very similar to the complex Fourier series just discussed. We summarize these properties for Fourier cosine series. If derivatives of $f(x)$ of order $p = 0, 1, \cdots, n-1$ are continuous for $0 < x < \pi$ while $f^{(p)}(0) = f^{(p)}(\pi) = 0$ for all *odd* $p < n$ and $f^{(n)}(x)$ is integrable, then the Fourier cosine coefficients given by (3.7) satisfy

(3.20) $$a_n \ll 1/k^n, \qquad k \to \infty,$$

as may be proven by integration by parts.

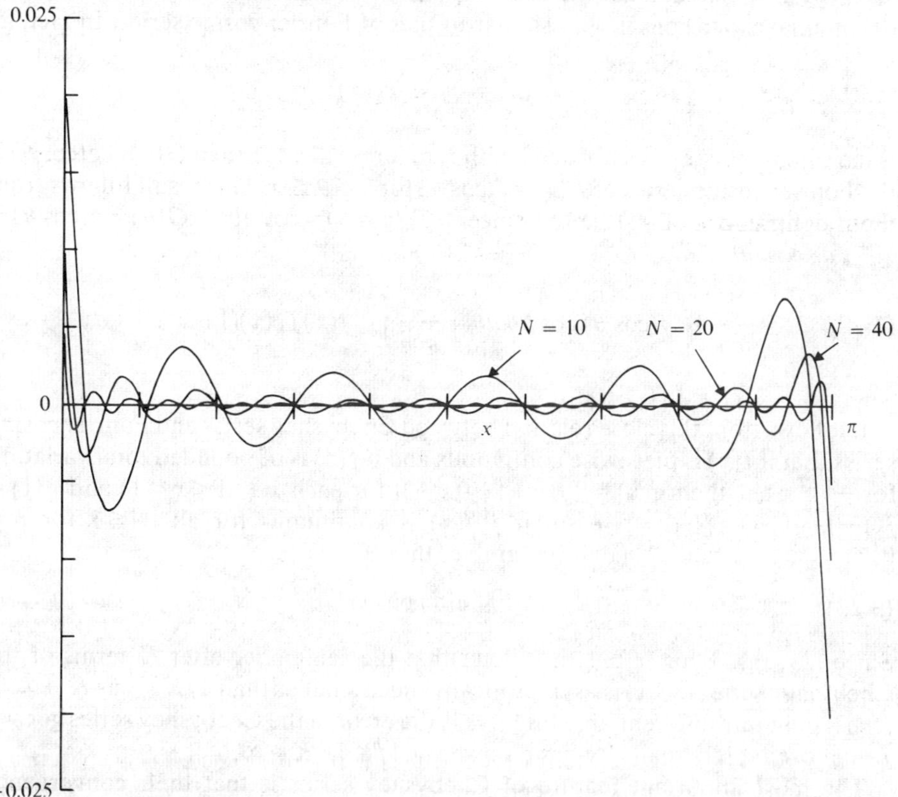

FIG. 3.4. *A plot of the error* $\sum_{n=0}^{N} a_n \cos nx - x/\pi$ *in the Fourier cosine series expansion of* x/π *for* $N = 10, 20, 40$. *Observe that there is no Gibbs phenomenon; the error decreases like* $1/N$ *near* $x = 0$ *and* $x = \pi$. *Also, observe that the error decreases like* $1/N^2$ *for fixed* x *in the interior of the interval* $0 < x < \pi$.

Thus, if $f(x)$ is infinitely differentiable for $0 \leqq x \leqq \pi$ and $f^{(2p+1)}(0) = f^{(2p+1)}(\pi) = 0$ for $p = 0, 1, \cdots$, then the Fourier cosine coefficients a_k approach zero more rapidly than any power of $1/k$ as $k \to +\infty$. In other words, if $f(x)$ is infinitely differentiable on $-\infty \leqq x \leqq \infty$, periodic with period 2π $[f(x+2\pi) = f(x)]$, and even $[f(x) = f(-x)]$, then the remainder after N terms of the Fourier cosine series (3.6) goes to zero more rapidly than any finite power of $1/N$ as $N \to \infty$.

To compare the convergence properties of Fourier sine and cosine series, we have plotted in Figs. 3.3 and 3.4 some results for the Fourier sine and cosine expansions, respectively, of the function x/π for $0 \leqq x \leqq \pi$. As discussed above, the Gibbs phenomenon in the sine series expansion is evident at $x = \pi$ (see Fig. 3.3). Observe that the error in the N term partial sum goes to zero like $1/N$ as $N \to \infty$ when x is fixed $0 \leqq x < \pi$. The Gibbs phenomenon near $x = \pi$ slows the convergence of the Fourier series for all x. In Fig. 3.4, we plot the error between the N term cosine series and x/π. Observe that as $N \to \infty$ the error goes to zero like $1/N^2$ for $0 < x < \pi$ and like $1/N$ when $x = O(1/N)$ as $N \to \infty$.

Chebyshev polynomial expansions. The convergence theory of Chebyshev polynomial expansions is very similar to that of Fourier cosine series. In fact, if

$$g(x) = \sum_{k=0}^{\infty} a_k T_k(x) \tag{3.21}$$

is the Chebyshev series associated with $f(x)$ for $-1 \leqq x \leqq 1$ then $G(\theta) = g(\cos\theta)$ is the Fourier cosine series of $F(\theta) = f(\cos\theta)$ for $0 \leqq \theta \leqq \pi$. This result follows from the definition of $T_n(x)$: since $T_n(\cos\theta) = \cos n\theta$, $G(\theta) = g(\cos\theta) = \sum_{k=0}^{\infty} a_n \cos n\theta$. Thus,

$$a_k = \frac{2}{\pi c_k} \int_0^{\pi} f(\cos\theta) \cos k\theta \, d\theta = \frac{2}{\pi c_k} \int_{-1}^{1} f(x) T_k(x)(1-x^2)^{-1/2} \, dx, \tag{3.22}$$

where $c_0 = 2$, $c_k = 1$ $(k > 0)$.

It follows from this close relation between Chebyshev series and Fourier cosine series that if $f(x)$ is piecewise continuous and if $f(x)$ is of bounded total variation for $-1 \leqq x \leqq 1$ then $g(x) = \frac{1}{2}[f(x+) + f(x-)]$ for each x $(-1 < x < 1)$ and $g(1) = f(1-)$, $g(-1) = f(-1+)$. Also, if $f^{(p)}(x)$ is continuous for all $|x| \leqq 1$ for $p = 0, 1, \cdots, n-1$, and $f^{(n)}(x)$ is integrable, then

$$a_k \ll 1/k^n, \qquad k \to \infty. \tag{3.23}$$

Since $|T_k(x)| \leqq 1$ for $|x| \leqq 1$, it follows that the remainder after K terms of the Chebyshev series (3.21) is asymptotically much smaller than $1/K^{n-1}$ as $K \to \infty$. If $f(x)$ is infinitely differentiable for $|x| \leqq 1$, the error in the Chebyshev series goes to zero more rapidly than any finite power of $1/K$ as $K \to \infty$.

The most important feature of Chebyshev series is that their convergence properties are not affected by the *values* of $f(x)$ or its derivatives at the boundaries $x = \pm 1$ but only by the *smoothness* of $f(x)$ and its derivatives throughout $-1 \leqq x \leqq 1$. In contrast, the Gibbs phenomenon shows that the rate of convergence of Fourier series depends on the values of f and its derivatives at the boundaries in addition to the smoothness of f in the interior of the interval. The reason for the absence of a Gibbs phenomenon for the Chebyshev series of $f(x)$ and its derivatives at $x=\pm 1$ is due to the fact that $F(\theta)=f(\cos\theta)$ satisfies $F^{(2p+1)}(0)=F^{(2p+1)}(\pi)=0$ provided only that all derivatives of $f(x)$ of order at most $2p+1$ exist at $x=\pm 1$.

An important consequence of the rapid convergence of Chebyshev polynomial expansions of smooth functions is that Chebyshev expansions may normally be differentiated termwise. Since

$$\frac{d^p}{dx^p}T_k(x)=O(k^{2p}), \qquad k\to\infty,$$

uniformly for $|x|\leqq 1$ [see (A.7)], if $a_k\to 0$ faster than any finite power of $1/k$ as $k\to\infty$ then (3.21) may be differentiated formwise:

$$\frac{d^p g}{dx^p}=\sum_{k=0}^{\infty} a_k\frac{d^pT_k(x)}{dx^p} \tag{3.24}$$

(as may be proven by an elementary uniform convergence argument).

While Chebyshev expansions do not exhibit the Gibbs phenomenon at the boundaries $x=\pm 1$, they do exhibit the phenomenon at any interior discontinuity of $f(x)$. In Fig. 3.5 we plot the partial sums of the Chebyshev expansions of the sign function sgn x:

$$\operatorname{sgn} x=\frac{4}{\pi}\sum_{n=0}^{\infty}(-1)^n\frac{T_{2n+1}(x)}{2n+1}. \tag{3.25}$$

Near $x=0$, a Gibbs phenomenon is observed; for fixed $x\neq 0$, the error after N terms is of order $1/N$. In general, the local structure of the partial sums $g_K(x)$ of Chebyshev series near a discontinuity of $f(x)$ at $x=x_0$ is, aside from a simple rescaling, given by (3.14):

$$g_K\left(x_0+\frac{z}{(K+\frac{1}{2})\sqrt{1-x_0^2}}\right)\sim\frac{1}{2}[f(x_0+)+f(x_0-)]+\frac{1}{\pi}[f(x_0+)-f(x_0-)]\operatorname{Si}(z), \qquad K\to\infty,$$

where $|x_0|<1$ and z is fixed. This equation is derived by a simple extension of the argument used to derive (3.14) [cf. (3.36) below for the explanation of the origin of the scale factor $1/\sqrt{1-x_0^2}$].

Rate of convergence of Sturm–Liouville eigenfunction expansions. Let us consider the expansion of a function $f(x)$ in terms of the eigenfunctions ϕ_n of a Sturm–Liouville problem: The eigenfunction $\phi_n(x)$ is a nonzero solution to

$$\frac{d}{dx}p(x)\frac{d\phi_n}{dx}+(\lambda_n w(x)-q(x))\phi_n(x)=0 \tag{3.26}$$

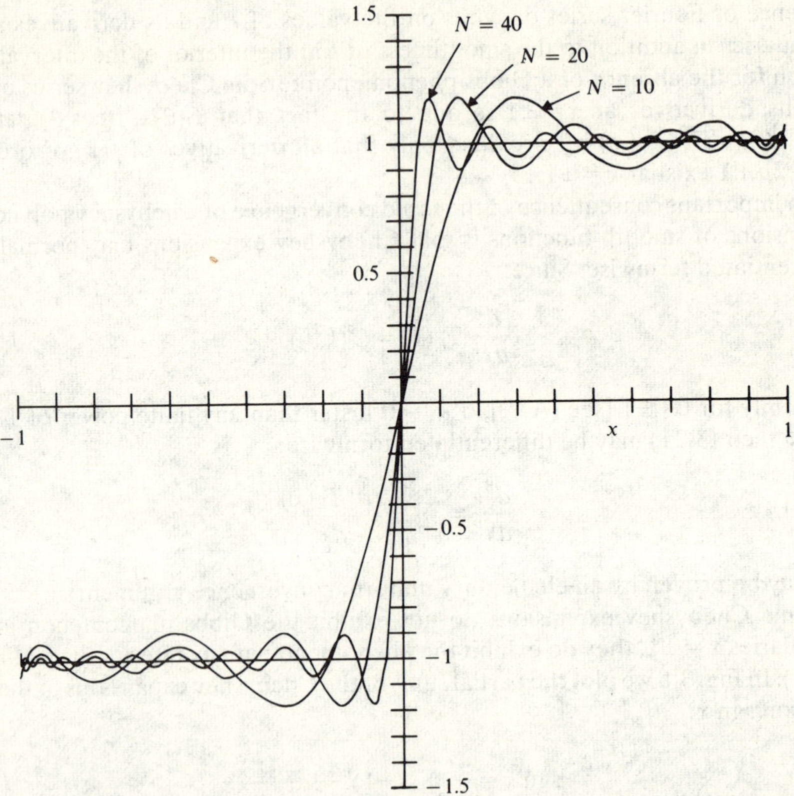

FIG. 3.5. *A plot of the partial sums of the Chebyshev series* (3.25) *for* sgn *x truncated after* $T_N(x)$ *for* $N=10, 20, 40$. *The function* sgn *x is also plotted. Observe the Gibbs phenomenon near* $x=0$. *Also observe that the series converges like* $1/N$ *for fixed* $x \neq 0$.

satisfying homogeneous boundary conditions. To be specific in our discussion we assume the boundary conditions $\phi_n(a)=\phi_n(b)=0$, although the analysis applies more generally. We assume that $p(x)\geqq 0$, $w(x)\geqq 0$, $q(x)\geqq 0$ for $a\leqq x\leqq b$. We will also assume that the eigenfunctions are normalized so that they satisfy

$$\int_a^b w(x)\phi_n(x)\phi_m(x)\,dx=\delta_{nm} \tag{3.27}$$

and that they form a complete set; the latter property follows if $\lambda_n\to\infty$ as $n\to\infty$ (see Courant and Hilbert (1953, p. 424)). The requirement that $\lambda_n\to\infty$ follows heuristically as follows: (3.26) suggests that $\phi_n(x)$ has a typical spatial scale of $1/\sqrt{\lambda_n}$, so the requirement that arbitrary $f(x)$ (with arbitrarily small spatial scale) be expansible in terms of $\{\phi_n\}$ implies that λ_n must grow unboundedly with n.

We wish to estimate the rate of convergence of the eigenfunction expansion

$$f(x)=\sum_{n=1}^{\infty} a_n\phi_n(x). \tag{3.28}$$

Using the orthonormality relation (3.27), the L_2-error after N terms is

$$\left[\int_a^b \left|f(x)-\sum_{n=1}^{N} a_n\phi_n(x)\right|^2 w(x)\,dx\right]^{1/2} = \left[\sum_{n=N+1}^{\infty} a_n^2\right]^{1/2}. \tag{3.29}$$

Thus, the L_2-error may be estimated by calculating the rate of decrease of a_n as $n\to\infty$.

Orthonormality of $\{\phi_n\}$ implies that

$$a_n = \int_a^b f(x)\phi_n(x)w(x)\,dx. \tag{3.30}$$

Substituting $w(x)\phi_n(x)$ from the Sturm–Liouville equation (3.26) gives

$$a_n = \frac{1}{\lambda_n}\int_a^b \left(-\frac{d}{dx}p(x)\frac{d\phi_n}{dx}+q(x)\phi_n\right)f(x)\,dx.$$

Integrating twice by parts, we obtain

$$a_n = \frac{1}{\lambda_n}p(x)[\phi_n(x)f'(x)-\phi_n'(x)f(x)]\Big|_{x=a}^{b} + \frac{1}{\lambda_n}\int_a^b h(x)\phi_n(x)w(x)\,dx, \tag{3.31}$$

where

$$h(x) = \left[-\frac{d}{dx}p(x)\frac{df}{dx}+q(x)f(x)\right]\Big/ w(x). \tag{3.32}$$

This integration by parts is justified if f is twice differentiable and h is square integrable with respect to w. Under these conditions and recalling that $\phi_n(a) = \phi_n(b) = 0$, we obtain

$$a_n = \frac{1}{\lambda_n}[p(a)\phi_n'(a)f(a)-p(b)\phi_n'(b)f(b)] + O\left(\frac{1}{\lambda_n}\right)$$

as $n\to\infty$, since $|\int_a^b h\phi_n w\,dx|^2 \leqq \int_a^b h^2 w\,dx \int_a^b \phi_n^2 w\,dx = O(1)$ as $n\to\infty$.

Nonsingular Sturm–Liouville problems. To proceed further we must distinguish between nonsingular and singular Sturm–Liouville problems: a problem is nonsingular if $p(x)>0$ and $w(x)>0$ throughout $a\leqq x\leqq b$. The important conclusion from (3.31)–(3.32) is that if the Sturm–Liouville problem is nonsingular and if $f(a)$ or $f(b)$ is nonzero then

$$a_n \sim \frac{1}{\lambda_n}[p(a)\phi_n'(a)f(a)-p(b)\phi_n'(b)f(b)], \qquad n\to\infty. \tag{3.33}$$

(Notice that if $\phi_n'(a)=0$, then $\phi_n(x)\equiv 0$ since (3.26) is a second-order differential equation and $p(x)\neq 0$.)

It is well known (Courant and Hilbert (1953)) that the asymptotic behavior of the eigenvalues and eigenfunctions of a *nonsingular* Sturm–Liouville problem are given by

$$\lambda_n \sim \left[n\pi \Big/ \int_a^b \sqrt{\frac{w}{p}}\, dx \right]^2, \qquad n \to \infty, \tag{3.34}$$

$$\phi_n(x) \sim A_n \sin\left(\sqrt{\lambda_n} \int_a^x \sqrt{\frac{w}{p}}\, dx \right), \qquad n \to \infty. \tag{3.35}$$

Using (3.34)–(3.35) in (3.33), we find that a_n behaves like $1/n$ as $n \to \infty$ if either $f(a) \neq 0$ or $f(b) \neq 0$. This behavior of a_n leads to the Gibbs phenomenon in the expansion (3.28) near those boundary points at which $f(a)$ or $f(b) \neq 0$. The asymptotic behavior (3.34)–(3.35) implies that this Gibbs phenomenon is asymptotically identical to that exhibited by Fourier sine series provided we use the stretched independent variable

$$X = \pi(x-a)\sqrt{w(a)/p(a)} \Big/ \int_a^b \sqrt{w(s)/p(s)}\, ds \tag{3.36}$$

near $x = a$ and a similarly stretched coordinate near $x = b$.

If $f(a) = f(b) = 0$, then $a_n \ll 1/n$ as $n \to \infty$. However, a further integration by parts in (3.31) shows that if the Sturm–Liouville problem is nonsingular and if $h(a)$ or $h(b) \neq 0$, then a_n behaves like $1/n^3$ as $n \to \infty$. In general, unless $f(x)$ satisfies an infinite number of very special conditions at $x = a$ and $x = b$, then a_n decays algebraically as $n \to \infty$.

These results on algebraic decay of errors in expansions based on nonsingular second-order eigenvalue problems generalize to higher-order eigenvalue problems. For example, as $n \to \infty$, the expansion coefficients in a_n in $f(x) = \sum_{n=0}^{\infty} a_n \phi_n(x)$, where $\{\phi_n(x)\}$ are the normalized "beam" functions

$$\phi_n'''' = \lambda_n \phi_n, \qquad \phi_n(\pm 1) = \phi_n'(\pm 1) = 0,$$

behave like $1/n$ if $f(\pm 1) \neq 0$ (implying a Gibbs phenomenon at the boundaries $x = \pm 1$), like $1/n^2$ if $f(\pm 1) = 0$ but $f'(\pm 1) \neq 0$, like $1/n^5$ if $f(\pm 1) = f'(\pm 1) = 0$ but $f''''(\pm 1) \neq 0$, and so on.

Singular Sturm–Liouville problems. If $p(a) = 0$ in (3.33) then it is not necessary to require that $f(a) = 0$ to achieve $a_n \ll \phi_n'/\lambda_n$ as $n \to \infty$. For this reason, expansions based on eigenfunctions of a Sturm–Liouville problem that is singular at $x = a$ do not normally exhibit the Gibbs phenomenon at $x = a$. Furthermore, if the argument that led to (3.33) can be repeated on $h(x)$ given by (3.32) (this is possible if p/w, p'/w, and q/w are bounded and all derivatives of f are square integrable with respect to w) then the boundary contribution to a_n from $x = a$ is smaller than ϕ_n'/λ_n^2 as $n \to \infty$. If there are also no boundary contributions from $x = b$ when the operations leading to (3.33) are repeated indefinitely (which is true if $p(b) = 0$), then a_n decreases more rapidly than any power of $1/\lambda_n$ as $n \to \infty$.

The important conclusion is that *eigenfunction expansions based on Sturm–Liouville problems that are singular at $x=a$ and at $x=b$ converge at a rate governed by the smoothness of the function being expanded not by any special boundary conditions satisfied by the function.*

Fourier–Bessel series. A Fourier–Bessel series of order 0 is obtained by choosing the expansion functions to be the eigenfunctions of the singular Sturm–Liouville problem

(3.37)
$$\frac{d}{dx}x\frac{d\phi_n}{dx}+\lambda_n x\phi_n=0, \qquad 0<x<1,$$
$$\phi_n(1)=0, \qquad \phi_n(0) \text{ finite.}$$

Here, $p(x)=w(x)=x$ in (3.26) so the problem is singular at $x=0$, but nonsingular at $x=1$. The eigenfunctions are

$$\phi_n(x)=J_0(j_{0n}x)$$

where J_0 is the Bessel function of order 0 and j_{0n} is its nth zero, $J_0(j_{0n})=0$. The eigenvalues $\lambda_n=j_{0n}$ satisfy

$$j_{0n}\sim(n-\tfrac{1}{4})\pi, \qquad n\to\infty.$$

The Fourier–Bessel expansion of a function $f(x)$ is given by

(3.38a)
$$f(x)=\sum_{n=1}^{\infty}a_nJ_0(j_{0n}x).$$

The expansion coefficients a_n are given by (3.30):

(3.38b)
$$a_n=\frac{2}{J_0'(j_{0n})^2}\int_0^1 tf(t)J_0(j_{0n}t)\,dt,$$

because

$$\int_0^1 tJ_0(j_{0n}t)^2\,dt=\tfrac{1}{2}J_0'(j_{0n})^2.$$

For example, the Fourier–Bessel expansion of $f(x)=1$ is

(3.39)
$$1=-\sum_{n=1}^{\infty}\frac{2}{j_{0n}J_0'(j_{0n})}J_0(j_{0n}x).$$

In Fig. 3.6 we plot the 10, 20, and 40 term partial sums of the series (3.39). There are three noteworthy features of this plot:

(i) At $x=1$ there is apparently a Gibbs phenomenon. In fact, it is easy to show that this Gibbs phenomenon has the same structure as that for Fourier sine series:

$$-\sum_{n=1}^{N}\frac{2}{j_{0n}J_0'(j_{0n})}J_0\left(j_{0n}-\frac{\pi z j_{0n}}{N+\frac{1}{2}}\right)\sim\frac{2}{\pi}\mathrm{Si}(z), \qquad N\to\infty.$$

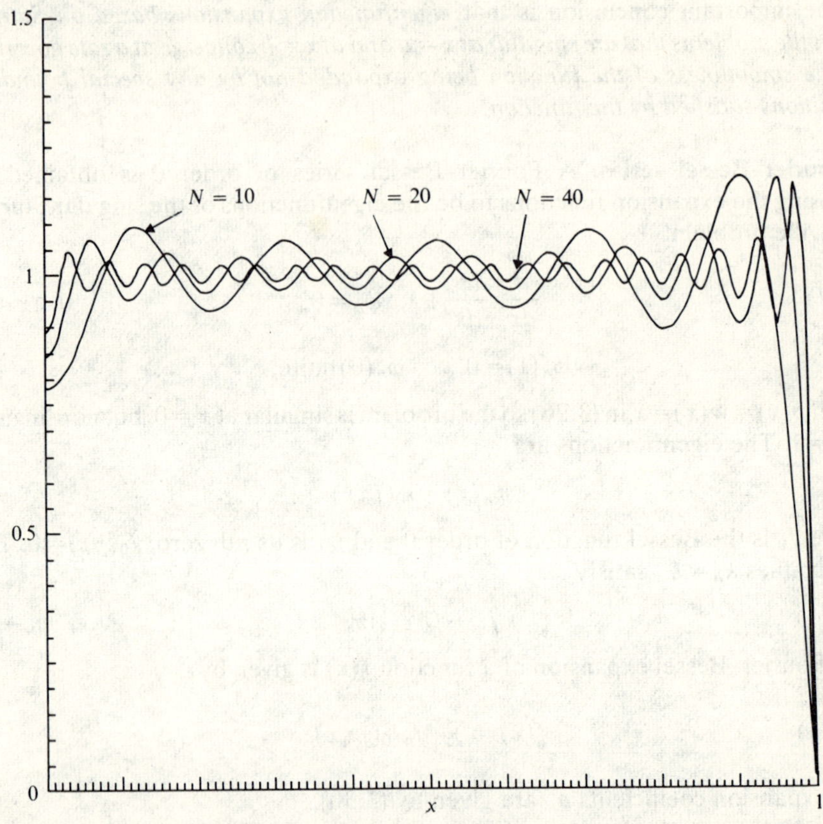

FIG. 3.6. *A plot of the partial sums of the Fourier–Bessel series expansion* (3.36) *of the function* 1 *truncated after* $N=10, 20, 40$ *terms. Observe the Gibbs phenomenon near* $x=1$. *Also observe that the series converges like* $1/N$ *for fixed x satisfying* $0<x<\pi$ *and like* $1/\sqrt{N}$ *for x near* 0.

This behavior is not too surprising because $J_0(t)\sim(2/(\pi t))^{1/2}\cos(t-\frac{1}{4}\pi)$ as $t\to+\infty$, so the large n behavior of (3.39) can be asymptotically approximated by that of Fourier series.

(ii) For fixed x satisfying $0<x<1$, the error after $N+1$ terms of (3.39) is

$$1+\sum_{n=0}^{N}\frac{2}{j_{0n}J_0'(j_{0n})}J_0(j_{0n}x)=O\left(\frac{1}{N}\right),\qquad N\to\infty.$$

In fact, the nth term of (3.39) has magnitude of order $1/n$ and oscillates in sign roughly every min $(1/x, 1/1-x)$ terms. The error in such an oscillating series is of order $1/N$ after N terms.

(iii) At $x=0$, the series converges (so there is no Gibbs phenomenon there) but the convergence is very slow and oscillatory. In fact, the error after N terms is of order $(-1)^{N+1}/\sqrt{N}$. This follows because $J_0(0)=1$ and

$$1+\sum_{n=0}^{N}\frac{2}{j_{0n}J_0'(j_{0n})}\sim\sqrt{2}\sum_{n=N+1}^{\infty}\frac{(-1)^n}{\sqrt{n}}\sim\frac{(-1)^{N+1}}{\sqrt{2N}},\qquad N\to\infty. \tag{3.40}$$

This slow rate of convergence near $x=0$ holds even though the eigenvalue problem is singular at $x=0$. There are two reasons for the slow convergence of Fourier–Bessel series near $x=0$. First, the Gibbs phenomenon at $x=1$ affects the rate of convergence throughout $0 \leqq x \leqq 1$. In fact, this is the sole source of the behavior (3.40). However, when $f'(x) \neq 0$, slow convergence near $x=0$ can also originate because $p(x)=w(x)=x$ gives $p'/w=1/x$ which is singular at $x=0$ so $h(x)$ given by (3.32) is singular at $x=0$ if $f'(0) \neq 0$.

Chebyshev series revisited. Chebyshev polynomials are the eigenfunctions of the singular Sturm–Liouville problem (3.26) with $p(x)=\sqrt{1-x^2}$, $w(x)=1/\sqrt{1-x^2}$, $q(x)=0$, $-1 \leqq x \leqq 1$, and the boundary conditions that $\phi_n(\pm 1)$ be finite. The eigenvalue corresponding to $T_n(x)$ is $\lambda_n = n^2$. Since $p/w = 1-x^2$ and $p'/w=-x$ are both finite for $|x| \leqq 1$, it follows that the argument leading from (3.30) to (3.33) can be repeated on $h(x)$ given by (3.32) so long as $f(x)$ is sufficiently differentiable. Therefore, the Chebyshev series expansion of an infinitely differentiable function converges faster than any power of $1/n$ as $n \to \infty$, as shown following (3.23) by a different method.

To illustrate the convergence properties of Chebyshev series expansions, we study the rate of convergence of the series

$$\text{(3.41)} \qquad \sin M\pi(x+a) = 2 \sum_{n=0}^{\infty} \frac{1}{c_n} J_n(M\pi) \sin\left(M\pi a + \tfrac{1}{2}n\pi\right) T_n(x), \qquad |x| \leqq 1.$$

Since $J_n(M\pi) \to 0$ exponentially fast as n increases beyond $M\pi$, it follows that (3.41) starts converging very rapidly when more than $M\pi$ terms are included (see Fig. 3.7). This result leads to an heuristic rule for the resolution requirements of Chebyshev expansions. Since $\sin M\pi(x+a)$ has M complete wavelengths lying within the interval $|x| \leqq 1$, we argue that Chebyshev expansions converge rapidly when at least *π polynomials are retained per wavelength.* In general, we expect that the Chebyshev expansion of a function that oscillates over a distance λ converges rapidly if at least $2\pi/\lambda$ polynomials are retained. Fewer polynomials are required if the region of rapid change of the function occurs only at the boundary (see below).

The Chebyshev polynomial expansion of a function $f(z)$ that is analytic in a region of the complex-z plane that includes the interval $-1 \leqq z \leqq 1$ converges at least exponentially fast as $n \to \infty$. If $f(z)$ has singularities in the finite-z plane then

$$\text{(3.42)} \qquad \limsup_{k \to \infty} |a_k|^{1/k} = \frac{1}{R}$$

where R is the sum of the semi-major and semi-minor axes of the largest ellipse with foci at $z=\pm 1$ within which $f(z)$ has no singularities. Thus, the L_2-error (3.29) after N terms of the Chebyshev expansions decays to 0 roughly like R^{-N} as $N \to \infty$.

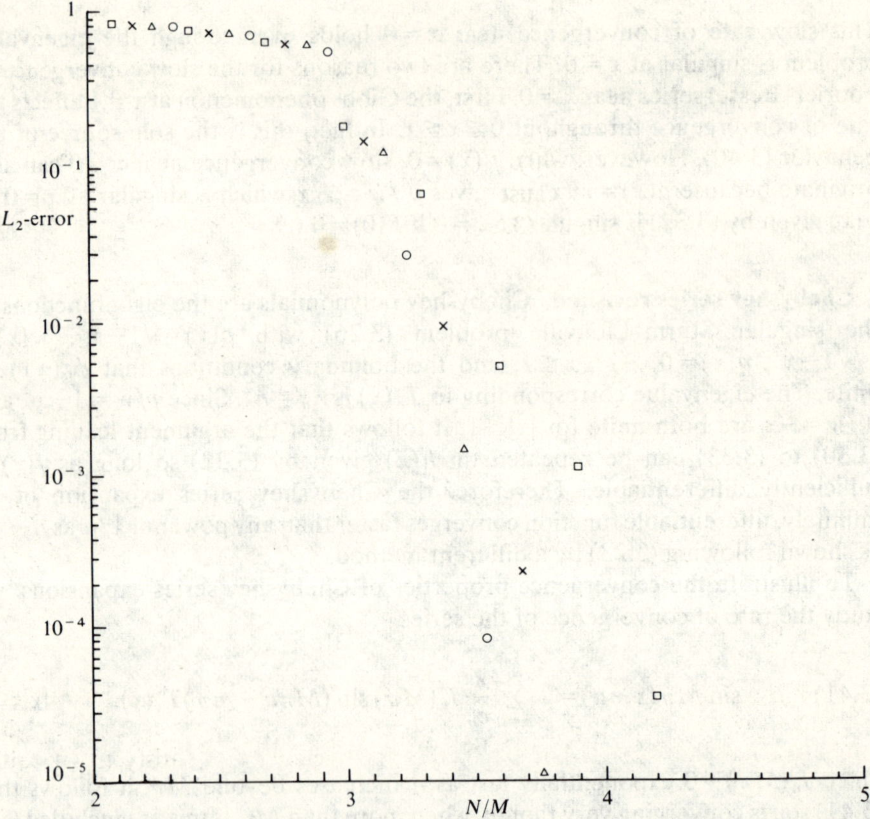

FIG. 3.7. *A plot of the* L_2*-error in the Chebyshev series expansion* (3.41) *of* $\sin(M\pi x)$ *truncated after* $T_N(x)$ *versus* N/M. *The various symbols represent*: $\square\, M=10$; $\times\, M=20$; $\triangle\, M=30$; $\bigcirc\, M=40$. *Observe that the* L_2*-error approaches zero rapidly when* $N/M>\pi$.

To prove (3.42), we note that

$$a_n=\frac{2}{\pi c_n}\int_{-1}^{1}\frac{f(z)T_n(z)}{\sqrt{1-z^2}}\,dz$$

(3.43)

$$=\frac{1}{\pi c_n}\int_C f(z)(1-z^2)^{-1/2}(z+\sqrt{z^2-1})^{-n}\,dz,$$

where C is any contour that encircles the interval $(-1,1)$ just once and does not enclose singularities of $f(z)$. Equation (3.43) follows because $2T_n(z)=(z+\sqrt{z^2-1})^{-n}+(z-\sqrt{z^2-1})^{-n}$, where we choose the branch of $\sqrt{z^2-1}$ satisfying $\sqrt{z^2-1}\sim z$ as $z\to\infty$. Since $(z+\sqrt{z^2-1})^{-n}\to 0$ as $z\to\infty$ with this choice of branch cut, we can expand the contour C to infinity by Cauchy's theorem and pick up the contributions from the singularities of $f(z)$. If the "nearest" singularity is a pole at $z=z_0$ with residue r (other singularities may be treated similarly), then

$$a_n\sim 2i\frac{r}{\sqrt{1-z_0^2}}(z_0+\sqrt{z_0^2-1})^{-n},\qquad n\to\infty.$$

To complete the justification of (3.42) we need only show that $|z_0+\sqrt{z_0^2-1}|=R$. Recall that an ellipse with foci at ± 1 satisfies $x^2/A^2+y^2/B^2=1$ with $A^2-B^2=1$. If z_0 lies on this ellipse, then setting $z_0=A\cos\theta+iB\sin\theta$, we have $z_0+\sqrt{z_0^2-1}=(A+B)e^{i\theta}=\mathrm{R}e^{i\theta}$.

Let us give an example of the behavior (3.42). The function $f(z)=\tanh(10z)$ has poles at $z=\pm i\pi/20$. Thus, $R=\pi/20+\sqrt{1+(\pi/20)^2}\doteq 1.16934$. The Chebyshev expansion coefficients of $f(z)$ satisfy $a_{2n}=0$ (because $f(z)$ is an odd function), while $a_1\doteq 1.2679$, $a_3\doteq -0.4089$, $a_5\doteq 0.2300$, and so on. The rms (L_2) error e_N (see (3.29)) obtained by truncating the series for $f(z)$ after $T_N(z)$ satisfies $(e_9/e_{11})\doteq(1.175)^2$, $e_{47}/e_{49}\doteq(1.16935)^2$, demonstrating (3.42) for this case. The error e_N is smaller than 0.01 for $N\geqq 25$, which again illustrates the result that roughly π polynomials per "wavelength" are required to resolve a function; the function $f(z)$ has a region of rapid change near $x=0$ of width roughly 0.2.

If $f(z)$ is entire, $R=\infty$ in (3.42) so its Chebyshev expansion coefficients decay faster than exponentially. More precisely, the method of steepest descents applied to (3.43) gives the following result: if $f(z)$ is entire and $f(z)=O(|z|^\beta\exp|z|^\alpha)$ as $z\to\infty$, then

$$\text{(3.44)}\qquad \limsup_{n\to\infty}(\ln|z_n|)/(n\ln n)=-1/\alpha.$$

For example, $\sin M\pi(z+a)$ is entire with $\alpha=1$ while its Chebyshev coefficients in (3.41) satisfy $a_n=O((M\pi)^n/n!)$ as $n\to\infty$, in agreement with (3.44). Also, a polynomial has Chebyshev coefficients that satisfy (3.44) with $\alpha=0$.

Finally, we remark that the Chebyshev series expansion (3.21)–(3.22) of an arbitrary function $g(x)$ has a maximum pointwise error that does not differ drastically from the smallest possible maximum pointwise error of any Nth degree polynomial, the so-called minimax error. In fact, the maximum pointwise error of the Chebyshev series (3.21) truncated after $T_N(x)$ is at most $4(1+\pi^{-2}\ln N)$ times larger than the minimax error (Rivlin (1969)). Since $4(1+\pi^{-2}\ln N)<10$ for $N<2{,}688{,}000$, the Chebyshev series is within a decimal place of the minimax approximation for all such polynomial approximations.

Legendre series. Legendre polynomials are the eigenfunctions of the singular Sturm–Liouville problem (3.26) with $p(x)=1-x^2$, $q(x)=0$, $w(x)=1$ for $-1\leqq x\leqq 1$ and the boundary conditions that $\phi_n(\pm 1)$ be finite. The nth eigenvalue is $\lambda_n=n(n+1)$ and its eigenfunction is $\phi_n(x)=P_n(x)$, the Legendre polynomial of degree n. Since $p/w=1-x^2$ and $p'/w=-2x$ are both finite for $|x|\leqq 1$, it follows that the Legendre series expansion of infinitely differentiable functions converges faster than algebraically.

To illustrate the convergence properties of Legendre series, we study the convergence of the series

$$\text{(3.45)}\quad \sin M\pi(x+a)=\frac{1}{\sqrt{2M}}\sum_{n=0}^{\infty}(2n+1)J_{n+1/2}(M\pi)\sin\left(M\pi a+\tfrac{1}{2}n\pi\right)P_n(x).$$

Since the expansion coefficients in (3.45) vanish rapidly as n increases beyond $M\pi$, we conclude that Legendre polynomial expansions of smooth functions converge rapidly provided that at least *π polynomials are retained per wavelength* (see Fig. 3.8).

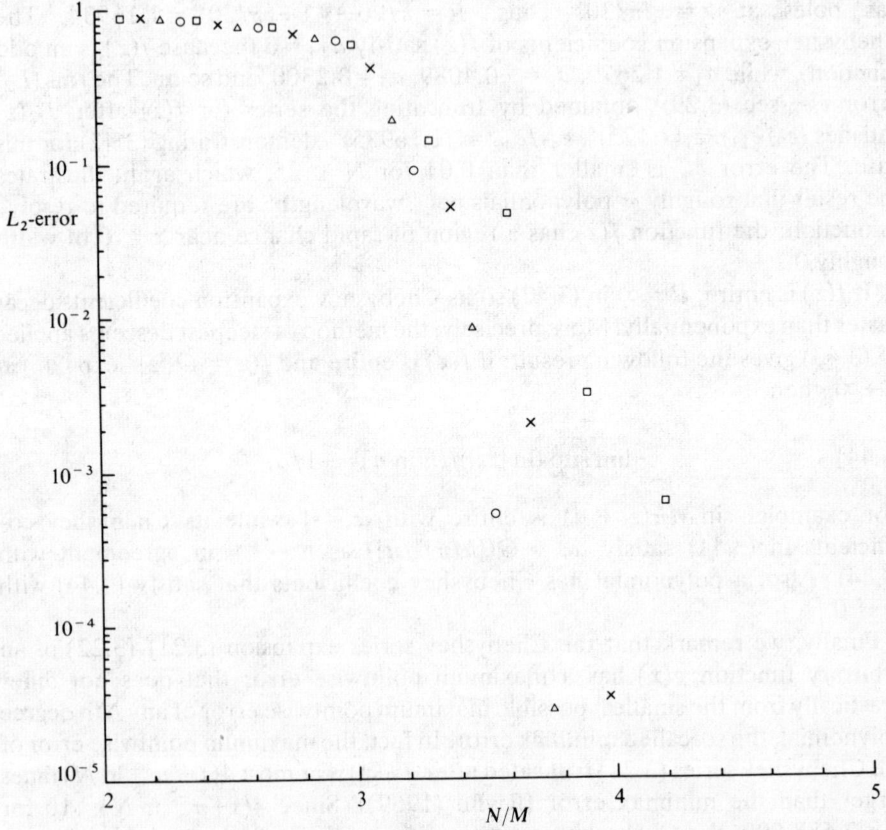

FIG. 3.8. *A plot of the L_2-error in the Legendre series expansion* (3.45) *of* $\sin(M\pi x)$ *truncated after* $P_N(x)$ *versus* N/M. *The various symbols represent*: □ $M=10$; × $M=20$; △ $M=30$; ○ $M=40$. *Observe that the L_2-error approaches zero rapidly when* $N/M>\pi$.

When a discontinuous function is expanded in Legendre series, the rate of convergence is no longer faster than algebraic. In the neighborhood of a discontinuity, a Gibbs phenomenon occurs whose local structure is the same as that for Fourier series with a suitably stretched coordinate. For example, the Legendre series expansion of the sign function sgn x is

$$\text{sgn}\, x = \sum_{n=0}^{\infty} \frac{(-1)^n(4n+3)(2n)!}{2^{2n+1}(n+1)!n!} P_{2n+1}(x). \tag{3.46}$$

The partial sums of this series are plotted in Fig. 3.9. Three features are noteworthy:

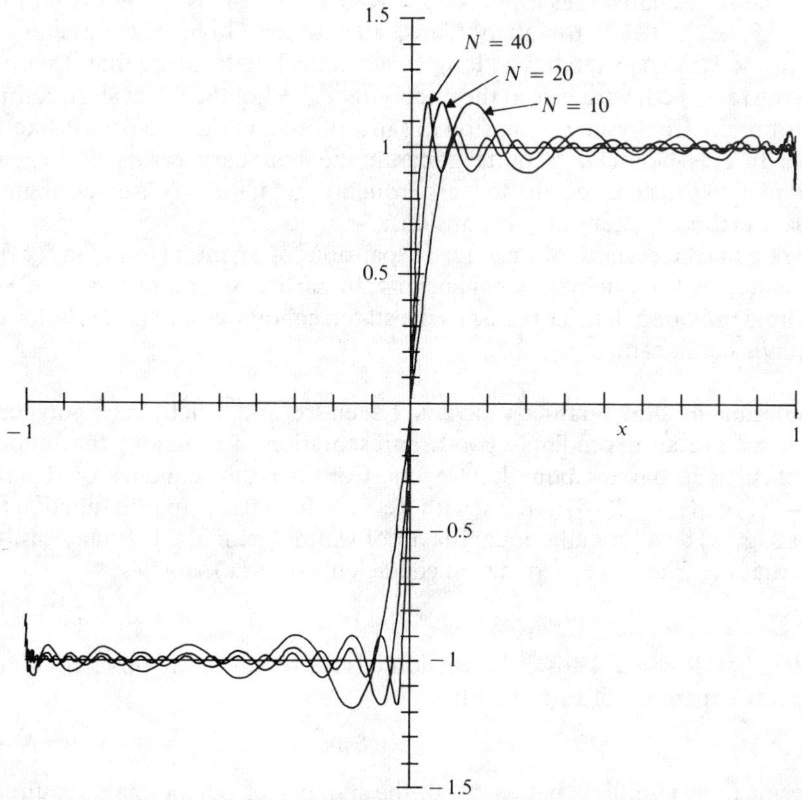

FIG. 3.9. *A plot of the partial sums of the Legendre series* (3.46) *for* sgn x *truncated after* $P_N(x)$ *for* $N=10, 20, 40$. *The function* sgn x *is also plotted. Observe the Gibbs phenomenon near* $x=0$. *Also observe that the series converges like* $1/N$ *for fixed* x, $x \neq 0, \pm 1$, *and that the series converges like* $1/\sqrt{N}$ *near* $x = \pm 1$.

(i) The Gibbs phenomenon near $x=0$ has the same structure as that for Fourier series.

(ii) The error after N terms behaves like $1/N$ for $|x|<1$, $x \neq 0$. This follows because the $(2n+1)$st Legendre coefficient in (3.46) satisfies

$$a_n = (-1)^n \frac{(4n+3)(2n)!}{2^{2n+1}(n+1)!n!} = O\left(\frac{1}{\sqrt{n}}\right), \qquad n \to \infty, \tag{3.47}$$

while $P_n(x)$ satisfies

$$P_n(x) = O\left(\frac{1}{\sqrt{n}}\right), \qquad n \to \infty,$$

for $|x|<1$ (see (A.29)). The series (3.46) is an alternating series if x is fixed away from zero so the error after N terms is at most of order $a_N P_N = O(1/N)$.

(iii) The series converges only like $1/\sqrt{N}$ at $x=\pm 1$. This follows from (3.47) because $P_n(\pm 1)=(\pm 1)^n$ for all n. Thus, an interior Gibbs phenomenon in a Legendre series expansion has a "long-range" effect in the sense that it seriously affects the rate of convergence at the endpoints $x=\pm 1$ of the interval. In contrast, the error of the Chebyshev expansion of sgn x plotted in Fig. 3.5 decays like $1/N$ at $x=\pm 1$. This behavior is quite general; the boundary errors of Legendre polynomial expansions decay to zero roughly a factor $\sqrt{N}$ slower than the boundary errors of Chebyshev expansions.

The rate of convergence of Legendre expansions of a general function $f(x)$ may be estimated as for Chebyshev expansions. In particular, the results (3.42) and (3.44) hold provided that $f(x)$ satisfies the stated conditions and (3.23) holds with only minor modifications.

Resolution of thin boundary layers. Legendre and Chebyshev polynomial expansions give an exceedingly good representation of functions that undergo rapid changes in narrow boundary layers. Consider the sequence of functions $g_\delta(x)=f(x)\exp[(x-1)/\delta]$ as $\delta\to 0$ with Re $\delta>0$ for a fixed smooth function $f(x)$. As $\delta\to 0$, $g_\delta(x)$ develops a boundary layer of width δ near $x=1$. It may easily be shown that the Chebyshev expansion coefficients of $g_\delta(x)$ satisfy

(3.48) $$a_n\sim(2\delta/\pi)^{1/2}f(1)\,e^{-n^2\delta/2},\qquad n\to\infty,\quad \delta n^2=O(1),$$

provided that Re $\delta>0$. Thus, if N polynomials are retained, the rms error ε in the Chebyshev expansion of $g_\delta(x)$ satisfies

(3.49) $$\ln\varepsilon\sim-\tfrac{1}{2}(\mathrm{Re}\,\delta)N^2,\qquad N\to\infty.$$

The result (3.49) implies that as $\delta\to 0$, the number of polynomials required to reach a specified error bound increases only as $1/\sqrt{\delta}$, in contrast to a uniform grid representation of $g_\delta(x)$ that would require order $1/\delta$ grid points in the interval $|x|\leqq 1$. In fact, to achieve 1% maximum pointwise error in boundary layers of thickness δ at the ends of the interval $-1\leqq x\leqq 1$, it is necessary to retain only

(3.50) $$N\sim 3/\sqrt{\mathrm{Re}\,\delta}$$

polynomials as $\delta\to 0$.

Heuristically, the reason that Chebyshev expansions represent boundary layers so well is that the extrema of $T_n(x)$ occur at $x_j=\cos\pi j/n$ for $j=0,1,\cdots,n$. Since $x_0-x_1\sim\pi^2/(2n^2)$ and $x_{n-1}-x_n\sim\pi^2/(2n^2)$ as $n\to\infty$, it follows that these polynomials can resolve changes over distances of order n^{-2}.

The convergence properties of Legendre polynomial expansions of boundary-layer functions are similar to those of Chebyshev expansions. In particular, (3.49) and (3.50) are both still valid. In Fig. 3.10 we compare the spatial distribution of the errors in Chebyshev and Legendre polynomial expansions of the function $g(x)=e^{100(x-1)}$, which has a narrow boundary layer of width 1/100 near $x=1$. Apparently for x away from the boundaries $x=\pm 1$, the Legendre expansion has somewhat smaller errors, while near $x=\pm 1$ the Chebyshev expansion has smaller errors.

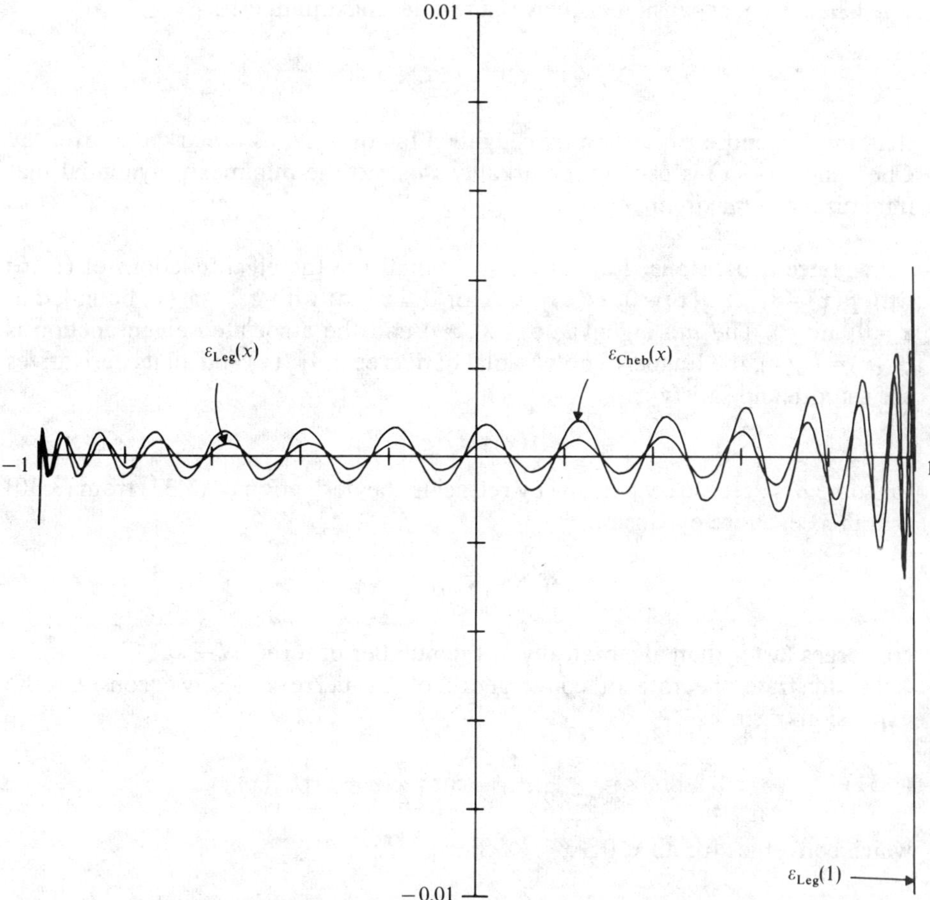

FIG. 3.10. *A plot of the errors* $\varepsilon(x) = Q_N(x) - \exp[100(x-1)]$ *in the Chebyshev and Legendre polynomial expansions of* $\exp[100(x-1)]$ of degree $N = 29$. *Observe that the Legendre expansion has smaller errors away from the boundaries* $x = \pm 1$ *but that the Chebyshev expansion has a maximum pointwise error that is about* 3 *times smaller than the Legendre expansion. Here we denote the Chebyshev polynomial approximation errors by* $\varepsilon_{\text{Cheb}}(x)$ *and the Legendre approximation errors by* $\varepsilon_{\text{Leg}}(x)$.

The Legendre expansion gives the polynomial $Q_N(x)$ of degree N that minimizes

$$\int_{-1}^{1} |g(x) - Q_N(x)|^2\, dx$$

while the Chebyshev expansion gives that Q_N that minimizes

$$\int_{-1}^{1} |g(x) - Q_N(x)|^2 (1-x^2)^{-1/2}\, dx.$$

The Chebyshev expansion also gives a smaller maximum error

$$\max_{|x|\leqq 1} |g(x) - Q_N(x)|$$

than the Legendre expansion by roughly a factor $\frac{1}{2}\sqrt{N}$; as remarked above, the Chebyshev $Q_N(x)$ is usually remarkably close to the minimax polynomial that minimizes the maximum error.

Laguerre expansions. Laguerre polynomials are the eigenfunctions of (3.26) with $p(x) = xe^{-x}$, $q(x) = 0$, $w(x) = e^{-x}$ for $0 \leqq x < \infty$ with $e^{-x/2}\phi_n(x)$ bounded at $x = 0$ and ∞. The nth eigenvalue is $\lambda_n = n$ and the associated eigenfunction is $\phi_n(x) = L_n(x)$, the Laguerre polynomial of degree n. If $f(x)$ and all its derivatives are smooth and satisfy

$$f(x) = O(e^{\alpha x}), \qquad x \to \infty,$$

for some $\alpha < \frac{1}{2}$, it is easy to show by retracing the derivation of (3.33) from (3.30) that the Legendre expansion

$$f(x) = \sum_{n=0}^{\infty} a_n L_n(x)$$

converges faster than algebraically as the number of terms $N \to \infty$.

To illustrate the rate of convergence of Laguerre series, we consider the expansion of $\sin x$:

$$\sin x = \sum_{n=0}^{\infty} \frac{1}{2^{(n+1)/2}} \cos\left[\frac{\pi}{4}(n+1)\right] L_n(x) \tag{3.51}$$

which converges for all x, $0 \leqq x < \infty$. Since

$$L_n(x) \sim \frac{1}{\sqrt{\pi}} e^{x/2} x^{-1/4} n^{-1/4} \cos\left[2\sqrt{nx} - \tfrac{1}{4}\pi\right], \qquad n \to \infty$$

[see Erdélyi et al. (1953, p. 200)], it follows that if $N \gg x$, then the error after N terms in (3.51) is roughly

$$\frac{e^{x/2}}{2^{N/2}(Nx)^{1/4}}.$$

This error is small only if $N \ln 2 > x$ or $N \gtrsim 1.44x$. Since the wavelength of $\sin x$ is 2π, Laguerre expansions require approximately *9.06 polynomials per wavelength* to achieve high accuracy. (This figure may be reduced to about 6.53 polynomials per wavelength by using the modified Laguerre expansion $\sum a_n L_n(x) e^{-\alpha x}$ and optimizing the choice of α.) Thus, Laguerre expansions require many more terms to resolve a function of given complexity than do either Chebyshev or Legendre expansions. The reason is that significant weight is given to $x \to +\infty$ in the Laguerre series where $\sin x$ has an essential singularity.

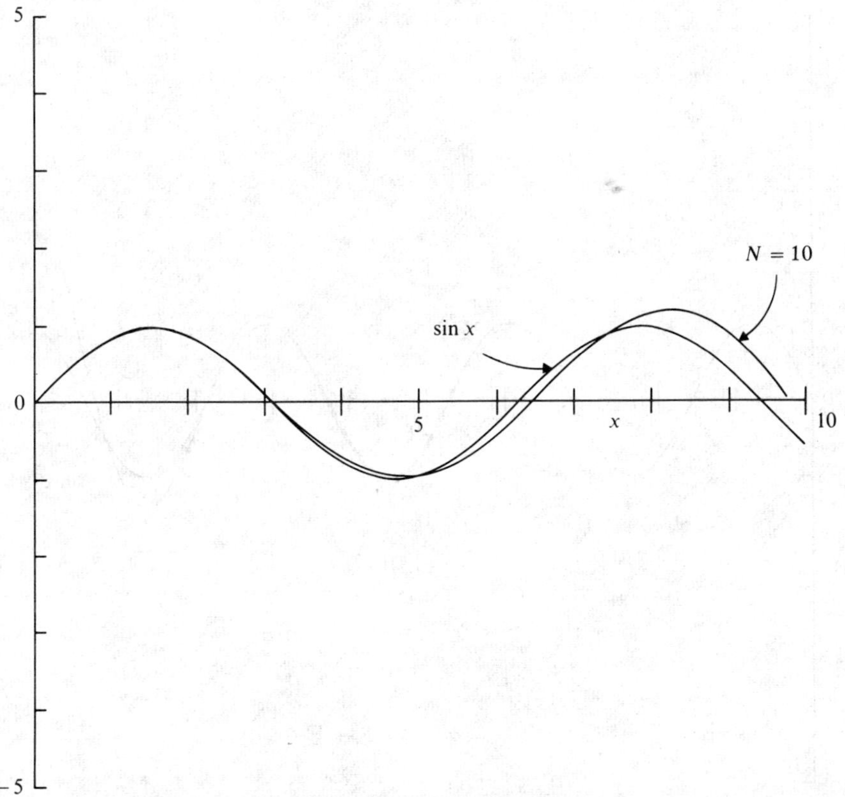

FIG. 3.11. *A plot of* sin *x and the partial sum of the Laguerre series expansion of* sin *x in* (3.51) *truncated after the Laguerre polynomial* $L_N(x)$ *with* $N=10$. *Observe that the accuracy of the approximation decreases as x increases and that roughly one wavelength of* sin *x is approximated accurately by this truncation of* (3.51).

In Figs. 3.11–3.13, we plot the partial sums of (3.51) with $N=10$, 20, and 40 terms. Observe that the number of wavelengths of sin x represented accurately by (3.51) is roughly $N/9$.

Hermite expansions. Hermite polynomials satisfy (3.26) with $p=e^{-x^2}$, $q(x)=0$, $w(x)=e^{-x^2}$ for $-\infty<x<\infty$, $\phi_n(x)\,e^{-x^2/2}$ bounded as $|x|\to\infty$. The Hermite polynomial $H_n(x)$ of degree n is associated with the eigenvalue $\lambda_n=2n$. If $f(x)$ and all its derivatives satisfy

$$f(x)=O(e^{\alpha x^2}),\qquad |x|\to\infty,$$

for some $\alpha<\frac{1}{2}$, then the Hermite expansion

$$f(x)=\sum_{n=0}^{\infty} a_n H_n(x)$$

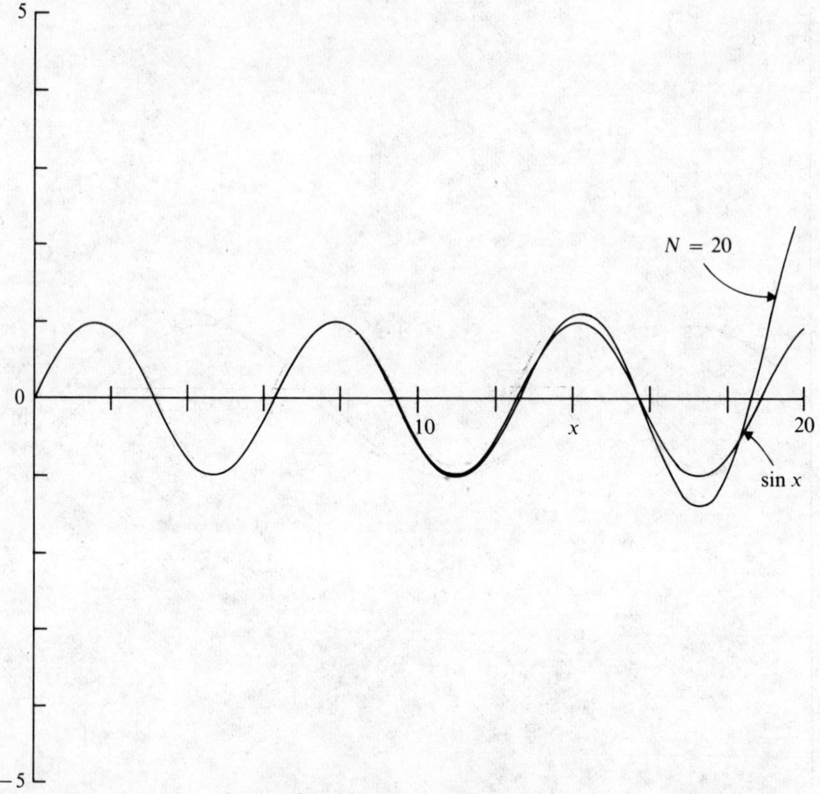

FIG. 3.12. *Same as Fig.* 3.11 *except that the Laguerre series* (3.51) *is truncated after* $L_N(x)$ *with* $N=20$. *Observe that the approximation is accurate for roughly* 2 *wavelengths of* sin x.

converges faster than algebraically as the number of terms $N \to \infty$. This is proved by retracing the steps leading from (3.30) to (3.33).

To study the rate of convergence of Hermite series, we consider the expansion of $\sin x$:

$$\sin x = \sum_{n=0}^{\infty} \frac{1}{2^{2n+1}(2n+1)!} H_{2n+1}(x). \tag{3.52}$$

Since the asymptotic behavior of $H_n(x)$ is given by (Erdélyi, et al. (1953, p. 201))

$$H_n(x) \sim e^{x^2/2} \frac{n!}{(\frac{1}{2}n)!} \cos(\sqrt{2n+1}x - \tfrac{1}{2}n\pi)$$

as $n \to \infty$ for x fixed, it follows that the error after N terms of (3.52) goes to zero rapidly at x only if $N \gtrsim x^2/\log x$. This result is very bad; to resolve M wavelengths of $\sin x$ requires nearly M^2 Hermite polynomials! (By expanding in the series $\sum a_n H_n(x) e^{-\alpha x^2}$ and optimizing the choice of α, it is possible to reduce the number of required Hermite polynomials to about $\frac{5}{2}\pi \doteq 7.85$ per wavelength, but this is still quite poor.)

Because of the poor resolution properties of Laguerre and Hermite polynomials the authors doubt they will be of much practical value in applications of spectral methods.

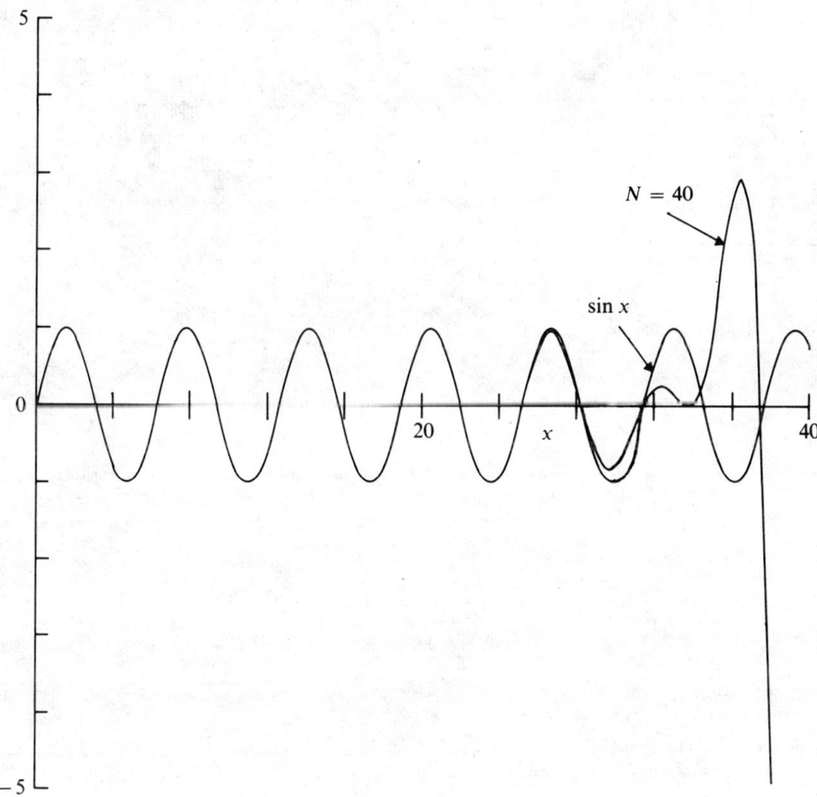

FIG. 3.13. *Same as Fig.* 3.12 *except that the Laguerre series* (3.51) *is truncated after* $L_N(x)$ *with* $N=40$. *Observe that the approximation is accurate for roughly* $4\frac{1}{2}$ *wavelengths of* $\sin x$.

SECTION 4

Review of Convergence Theory

The fundamental problem of the numerical analysis of initial value problems is to find conditions under which $u_N(x, t)$ converges to $u(x, t)$ as $N \to \infty$ for some time interval $0 \leqq t \leqq T$ and to estimate the error $\|u - u_N\|$. The principal result is the Lax–Richtmyer equivalence theorem which states that *stability* is equivalent to *convergence* for *consistent* approximations to well-posed linear problems. The terms stable, convergent, and consistent relate to technical properties of the approximation scheme which are defined below.

An approximation scheme (2.5)–(2.6) is *stable* if

$$\|e^{L_N t}\| \leqq K(t) \tag{4.1}$$

for all N where $K(t)$ is a finite function of t. Here the operator norm is defined by

$$\|e^{L_N t}\| = \max_{u \in \mathcal{H}} \frac{\|e^{L_N t} u\|}{\|u\|}.$$

An approximation scheme is *convergent* if

$$\|u(t) - u_N(t)\| \to 0 \quad \text{as } N \to \infty$$

for all t in the interval $0 \leqq t \leqq T$ and *all* $u(0) \in \mathcal{H}$ and $f(t) \in \mathcal{H}$. Finally, an approximation scheme is *consistent* if

$$\begin{aligned} &\|Lu - L_N u\| \to 0, \\ &\|u - P_N u\| \to 0 \end{aligned} \tag{4.2}$$

as $N \to \infty$ for all u in a dense subspace of $\mathcal{H}$.

The classical Lax–Richtmyer equivalence theorem relating the above definitions states that "a consistent approximation to a well-posed linear problem is stable if and only if it is convergent." In this monograph we are confronted with some subtleties regarding the notions of stability and convergence. Because a precise understanding of the ideas of stability and convergence is important to the theory of algebraic stability given in § 5, we outline here the proof of the equivalence theorem.

Proof of the equivalence theorem. To show that stability implies convergence we use (2.1) and (2.5) to obtain

$$\frac{\partial(u - u_N)}{\partial t} = L_N(u - u_N) + Lu - L_N u + f - f_N.$$

Thus,

$$u(t)-u_N(t)=e^{L_Nt}[u(0)-u_N(0)] \qquad (4.3)$$
$$+\int_0^t e^{L_N(t-s)}[Lu(s)-L_Nu(s)+f(s)-f_N(s)]\,ds.$$

Using (4.1) and (4.3) and the triangle inequality we obtain the estimate

$$\|u(t)-u_N(t)\|\leqq K(t)\|u(0)-u_N(0)\| \qquad (4.4)$$
$$+\int_0^t K(t-s)[\|Lu(s)-L_Nu(s)\|+\|f(s)-f_N(s)\|]\,ds.$$

Thus, if $u(t)$ belongs to the dense subspace of $\mathscr{H}$ satisfying (4.2) and if $f(t)$ belongs to the dense subspace of $\mathscr{H}$ satisfying $\|f-P_Nf\|\to 0$ as $N\to\infty$, then $\|u(t)-u_N(t)\|\to 0$ as $N\to\infty$. Since all solutions $u(t)$ of (2.1) can be approximated arbitrarily well by functions satisfying (4.2), the proof that stability implies convergence is completed.

Conversely, to show that convergence implies stability, we first observe that, for any $u\in\mathscr{H}$, $\|e^{L_Nt}u\|$ is bounded for all N and each fixed t. In fact, convergence implies

$$0\leqq\big|\|e^{L_Nt}u\|-\|e^{Lt}u\|\big|\leqq\|e^{L_Nt}u-e^{Lt}u\|\to 0, \qquad N\to\infty,$$

while well-posedness requires that $\|e^{Lt}u\|$ is finite. However, $\max_N\|e^{L_Nt}u\|$ may depend on u and on t, so stability is not yet proved. To complete the proof we use the fact that $\mathscr{H}$ is a Hilbert space. The principle of uniform boundedness (Richtmyer and Morton (1967)) implies that if $\|e^{L_Nt}u\|$ is bounded as $N\to\infty$ for each t and $u\in\mathscr{H}$, then $\|e^{L_Nt}\|$ is bounded as $N\to\infty$ for each t. This proves stability and completes the proof of the equivalence theorem.

Using the equivalence theorem, the study of the convergence of discrete approximations to the solutions of initial-value problems is reduced to the study of the stability of the discrete approximations, assuming the approximations are consistent. Thus, the development of conditions for the stability of families of finite-dimensional operators L_N is of primary interest in numerical analysis.

Von Neumann stability condition. The simplest condition for stability is due to von Neumann. Let us suppose that the Hilbert space $\mathscr{H}$ possesses the inner product $(\cdot,\cdot)$. Using the inner product, we define (neglecting the complications due to boundary conditions) the adjoint L^* of an operator L as that linear operator that satisfies $(u, Lv)=(L^*u, v)$ for all u, v in $\mathscr{H}$. For the finite dimensional approximation L_N, the matrix representation of L_N^* is the adjoint of the matrix representation of L_N (see § 2). The operator L_N is said to be a normal operator if L_N commutes with L_N^* so $L_NL_N^*=L_N^*L_N$.

The von Neumann stability condition is that stability of normal operators L_N is equivalent to the condition

$$\operatorname{Re}\lambda_N<C \qquad (4.5)$$

where λ_N is any of the eigenvalues of any of the operators L_N and C is a finite constant independent of N. To prove this, we note that if L_N is normal, then L_N and L_N^* as well as $\exp(L_N t)$ and $\exp(L_N^* t)$ are simultaneously diagonolizable. Therefore,

$$\|e^{L_N t}\|^2 = \max_{u \in \mathscr{H}} \frac{(u, e^{L_N^* t} e^{L_N t} u)}{(u, u)} = \max_{\lambda_N} e^{2(\mathrm{Re}\,\lambda_N)t}$$

where λ_N is an eigenvalue of L_N. Thus, the von Neumann condition (4.5) is equivalent to the stability definition (4.1) with $K(t) = \exp(Ct)$.

The von Neumann condition gives an operational technique for checking stability of normal approximations: compute the eigenvalues of L_N and check that the real parts of the eigenvalues are bounded from above.

EXAMPLE 4.1. *Symmetric hyperbolic system with periodic boundary conditions.* Let us apply the theory just discussed to the stability of difference approximations to the m-component symmetric hyperbolic system

$$\frac{\partial \mathbf{u}(x,t)}{\partial t} = A \frac{\partial \mathbf{u}(x,t)}{\partial x} \tag{4.6}$$

with periodic boundary conditions $\mathbf{u}(0, t) = \mathbf{u}(1, t)$. Here $\mathbf{u}$ is an m-component vector and A is a symmetric $m \times m$ matrix.

If we discretize in space using second-order centered differences, we obtain

$$\begin{gathered} \frac{\partial \mathbf{u}_j}{\partial t} = A \frac{\mathbf{u}_{j+1} - \mathbf{u}_{j-1}}{2\Delta x}, \qquad j = 1, 2, \cdots, N, \\ \mathbf{u}_0(t) = \mathbf{u}_N(t), \qquad \mathbf{u}_1(t) = \mathbf{u}_{N+1}(t), \end{gathered} \tag{4.7}$$

where $\mathbf{u}_k(t) = \mathbf{u}(k/N, t)$ and $\Delta x = 1/N$. The system (4.7) is equivalent to the system of mN equations

$$\frac{\partial \hat{u}}{\partial t} = B\hat{u} \tag{4.8a}$$

where $\hat{u}$ is the column vector of length mN whose transpose is $\hat{u}^T = (\mathbf{u}_1, \mathbf{u}_2, \cdots, \mathbf{u}_N)$. Here B is the $mN \times mN$ matrix given by the Kronecker product

$$B = A \otimes D, \tag{4.8b}$$

where A is the $m \times m$ matrix in (4.6) and D is the $N \times N$ matrix

$$D = \frac{1}{2\Delta x} \begin{pmatrix} 0 & 1 & 0 & 0 & \dots & 0 & -1 \\ -1 & 0 & 1 & 0 & \dots & 0 & 0 \\ 0 & -1 & 0 & 1 & \dots & 0 & 0 \\ \vdots & \vdots & \vdots & \vdots & & \vdots & \vdots \\ 0 & 0 & 0 & 0 & & 0 & 1 \\ 1 & 0 & 0 & 0 & & -1 & 0 \end{pmatrix}$$

D is anti-symmetric (so $D^* = -D$ and, hence, D is normal) so it has eigenvalues that are either 0 or pure imaginary. In fact, the eigenvalues of D are $i \sin(2\pi k\, \Delta x)/\Delta x$ for $k = 0, 1, \cdots, N-1$. Thus, the norm of $\exp(Bt)$ satisfies

$$\|\exp(Bt)\| = \max_{0 \leqq k < N} \|\exp[iAt \sin(2\pi k\, \Delta x)/\Delta x]\| = 1,$$

where we use the fact that A is symmetric so it has real eigenvalues.

Kreiss matrix theorem. If the approximate evolution operators L_N are not normal, conditions guaranteeing stability are much harder to obtain. The von Neumann condition (4.5) is still necessary for stability, but it is not sufficient to ensure stability. One important case in which stability conditions can be found is for the problem studied in Example 4.1 with A no longer symmetric. The appropriate generalization is to assume that the approximation L_N has the form $L_N = A \otimes D_N$ where A is a fixed $m \times m$ matrix (possibly not normal) and D_N is an N-dimensional normal matrix. It is easy to show that

$$\|\exp(L_N t)\| = \max_{\lambda_N} \|\exp(\lambda_N A t)\| \tag{4.9}$$

where λ_N is any one of the eigenvalues of D_N. A stability condition for (4.9) will be obtained below. To do this, we generalize (4.9) and seek conditions for the stability of a family of $m \times m$ matrices $A(\omega)$, where ω is an arbitrary parameter. That is, we seek conditions such that

$$\max_{\omega} \|\exp[A(\omega)t]\| \leqq K(t),$$

where $K(t)$ is a finite function of t. Once these general conditions are found, they can be specialized to give stability conditions for families of the form $\exp(L_N t)$ where $L_N = A \otimes D_N$ with D_N normal by simply choosing $A(\omega) = A\omega$ where ω is any one of the eigenvalues of any of the matrices D_N.

The basic result on the stability of families of $m \times m$ matrices is the Kreiss matrix theorem (Kreiss (1962)):

For any family $A(\omega)$ of $m \times m$ matrices, each of the following statements implies the next:

(i) There exist symmetric matrices $H(\omega)$ satisfying $H(\omega)A(\omega) + A^*(\omega)H(\omega) \leqq 0$ and $I \leqq H(\omega)$, $\|H(\omega)\| \leqq C$ for some constant C.

(ii) $\|\exp[A(\omega)t]\| \leqq C$ for all $t \geqq 0$.

(iii) $(\operatorname{Re} \lambda)\|(\lambda I - A(\omega))^{-1}\| \leqq C'$ for some constant C' and all λ satisfying $\operatorname{Re} \lambda > 0$.

(iv) There exist matrices $H(\omega)$ satisfying (i) with $\|H(\omega)\| \leqq K(m)C'$ where C' is the constant appearing in (iii) and $K(m)$ depends only on m and not only the family $A(\omega)$.

Observe that for a family of matrices $A(\omega)$ to satisfy the conditions of this theorem it is necessary that all the eigenvalues of all the matrices have nonpositive real parts. Otherwise there would be some ω and some eigenvector $\mathbf{u}$ satisfying $\|\exp[A(\omega)t]\mathbf{u}\| \to \infty$ as $t \to \infty$ violating (ii).

The most importnat relation implied by this theorem is the implication that (iii) implies (ii) with $C \leqq K(m)C'$. That is, for any $m \times m$ matrix A all of whose eigenvalues have nonpositive real parts

$$\|\exp(At)\| \leqq K'(m) \max_{\operatorname{Re}\lambda>0} (\operatorname{Re}\lambda)\|(\lambda I - A)^{-1}\| \tag{4.10}$$

where $K'(m)$ is a finite function of m.

An elementary proof of (4.10) has recently been given by Laptev (1975) and improved by C. McCarthy (private communication to G. Strang, (1975)). Laptev observes that if $\nu > 0$, then

$$e^{At} = \frac{1}{2\pi i}\int_{\nu-i\infty}^{\nu+i\infty} e^{\lambda t}(\lambda I - A)^{-1}\,d\lambda = \frac{e^{\nu t}}{2\pi}\int_{-\infty}^{\infty} e^{i\mu t}(\nu + i\mu - A)^{-1}\,d\mu, \tag{4.11}$$

as may be proved by shifting contours in the complex plane. Since each entry of $(\nu + i\mu - A)^{-1}$ is a rational function in μ of degree at most m, the derivatives of the real and imaginary parts of each entry can change sign at most 4 m times when μ increases from $-\infty$ to ∞. On any μ-interval, say $a \leqq \mu \leqq b$, where the real and imaginary parts of an entry in $(\nu + i\mu - A)^{-1}$ are monotonic, the second mean value theorem implies

$$\int_a^b \cos\mu t f(\mu)\,d\mu = f(a)\left[\frac{\sin(ct) - \sin(at)}{t}\right] + f(b)\left[\frac{\sin(bt) - \sin(ct)}{t}\right]$$
$$\leqq \frac{4}{t}\max_{\mu}|f(\mu)|,$$

for some c satisfying $a < c < b$ where $f(\mu)$ is the real or imaginary part of an entry in the matrix $(\nu + i\mu - A)^{-1}$. If we apply this kind of inequality to each element of the right side of (4.11), we obtain

$$\left|\int_{-\infty}^{\infty} e^{i\mu t}(\nu + i\mu - A)_{ij}^{-1}\,d\mu\right| \leqq \frac{64\,m}{t}\max_{\mu}\left|(\nu + i\mu - A)_{ij}^{-1}\right|, \quad 1 \leqq i, j \leqq m. \tag{4.12}$$

If the matrix norm has the property that $\max|B_{ij}| \leqq \|B\| \leqq m \max|B_{ij}|$ for all $m \times m$ matrices B, then (4.12) gives

$$\left\|\int_{-\infty}^{\infty} e^{i\mu t}(\nu + i\mu - A)^{-1}\,d\mu\right\| \leqq \frac{64\,m^2}{t}\max_{\mu}\left\|(\nu + i\mu - A)^{-1}\right\|. \tag{4.13}$$

Choosing $\nu = 1/t$ in (4.11–13) gives (4.10) with $K'(m) = 32\,em^2/\pi$.

There are three important matrix norms in which $\max|B_{ij}| \leqq \|B\| \leqq m \max|B_{ij}|$, namely the matrix norms induced by the L_1, L_2, and L_∞ vector norms.

This is shown using the relations

$$\|B\|_1 = \max_j \sum_{i=1}^m |B_{ij}|, \qquad \|B\|_\infty = \max_i \sum_{j=1}^m |B_{ij}|,$$

$$\|B\|_2 = \sup_{\substack{\|x\|_2=1\\ \|y\|_2=1}} \sum_{i=1}^m \sum_{j=1}^m B_{ij} x_i y_j,$$

which hold for all matrices B. In other norms it may not be true that $\max |B_{ij}| \leqq \|B\| \leqq m \max |B_{ij}|$, but the equivalence of all matrix norms implies that (4.10) must hold with some finite function $K'(m)$ of the dimension m.

The functions $K(m)$ appearing in statement (iv) of the Kreiss theorem and $K'(m)$ appearing in (4.10) need not be equal. It follows from the Kreiss theorem that the smallest function $K'(m)$ satisfying (4.10) satisfies $K'(m) \leqq K(m)$. Kreiss showed only that $K(m) = O(m^m)$ as $m \to \infty$; this is much too conservative. Miller and Strang (1965) showed that it is possible to choose $K(m)$ so that $K(m) = O(C^m)$ as $m \to \infty$ for some constant $C > 1$.

In the case of a normal family of matrices $A(\omega)$ the conditions of the Kreiss matrix theorem are trivially satisfied: if the eigenvalues of $A(\omega)$ have negative real parts, then $\|\exp[A(\omega)t]\| \leqq 1$ for all $t \geqq 0$ and ω.

Nonnormal approximations. The Kreiss matrix theorem applies to approximations of the form $L_N = A \otimes D_N$, where A is a fixed dimensional nonnormal matrix and D_N is an N-dimensional normal matrix. This type of operator L_N is commonly encountered in the solution of initial-value problems with periodic boundary conditions. On the other hand, nonperiodic boundary conditions usually lead to problems in which the nonnormality affects the N-dimensional matrix D_N. When finite-difference methods are used for such problems, the deviation of D_N from a normal operator is frequently "small".

EXAMPLE 4.2. *Nonnormality of a difference approximation to a mixed initial-boundary value problem.* A difference approximation to the mixed initial-boundary value problem

$$\frac{\partial u}{\partial t} + \frac{\partial u}{\partial x} = f(x, t), \qquad 0 \leqq x \leqq 1, \quad t > 0,$$

$$u(0, t) = 0,$$

$$u(x, 0) = g(x)$$

is given by

$$\frac{\partial u_j}{\partial t} + \frac{u_{j+1} - u_{j-1}}{2h} = f(jh, t), \qquad 1 \leqq j \leqq N, \tag{4.14}$$

where $u_j(t) = u(jh, t)$ and we set $u_0(t) = 0$ and $u_{N+1}(t) = 2u_N(t) - u_{N-1}(t)$. The latter condition is an extrapolation condition which ensures that (4.14) is a closed system of equations. This approximation has the matrix representation

$$L_N = -\frac{1}{2h}\begin{pmatrix} 0 & 1 & 0 & 0 & \cdot & \cdot & \cdot & \cdot & \cdot & \cdot & 0 & 0 & 0 \\ -1 & 0 & 1 & 0 & \cdot & \cdot & \cdot & \cdot & \cdot & \cdot & 0 & 0 & 0 \\ 0 & -1 & 0 & 1 & \cdot & \cdot & \cdot & \cdot & \cdot & \cdot & 0 & 0 & 0 \\ \vdots & \vdots & \vdots & \vdots & & & & & & & \vdots & \vdots & \vdots \\ 0 & 0 & 0 & 0 & \cdot & \cdot & & & \cdot & & -1 & 0 & 1 \\ 0 & 0 & 0 & 0 & \cdot & \cdot & & & & & 0 & -2 & 2 \end{pmatrix}.$$

The departure of L_N from a normal matrix is a matrix E of rank 1 with nonzero entries $E_{N,N-1}=-1$, $E_{N,N}=1$. For problems of this kind, extensions of von Neumann stability analysis, like that introduced by Godunov and Ryabenkii (1963) and extended by Kreiss (see Kreiss and Oliger (1973)), apply.

Unfortunately, the class of semi-discrete approximations investigated in this monograph include problems that cannot be easily analyzed either by straightforward von Neumann stability analysis or by the Godunov–Ryabenkii or Kreiss analysis. In contrast to the classical problems of the numerical analysis of difference methods for initial-value problems, spectral approximations L_N are frequently not even approximately normal.

SECTION 5

Algebraic Stability

In this section, we develop a theory of stability and convergence which generalizes the classical theory discussed in § 4. As will be shown by examples in §§ 6–8, this generalized stability theory is well suited to study the convergence of spectral methods.

A spectral approximation

$$\frac{\partial u_N}{\partial t} = L_N u_N + f_N \tag{5.1}$$

to the initial-value problem $u_t = Lu + f$ is called *algebraically stable* as $N \to \infty$ if

$$\|e^{L_N t}\| \leqq N^r N^{st} K(t) \tag{5.2}$$

for all sufficiently large N, where r, s, and $K(t)$ are finite for $0 \leqq t \leqq T$.

It may at first seem that the Lax–Richtmyer theorem shows that algebraically stable approximations cannot be convergent unless (5.2) holds with $r \leqq 0$, $s \leqq 0$. In fact, if we demand that the approximations converge for all $u(0)$ and $f(t)$ in the Hilbert space $\mathcal{H}$, this conclusion *is* correct. However, it is possible for approximations that satisfy (5.2) with $r > 0$ or $s > 0$ to converge on a dense subset of the Hilbert space in which the only functions for which convergence is not obtained are highly pathological. In fact, if $p = r + sT > 0$ but p is smaller than the order of the spatial truncation error of a particular solution $u(x, t)$, i.e.,

$$N^p \|Lu(t) - L_N u(t)\| \to 0, \qquad N \to \infty, \tag{5.3a}$$

$$N^p \|u(0) - u_N(0)\| \to 0, \qquad N \to \infty, \tag{5.3b}$$

$$N^p \|f(t) - f_N(t)\| \to 0, \qquad N \to \infty, \tag{5.3c}$$

for all $0 \leqq t \leqq T$, then (4.4) and (5.2) imply that

$$\|u(t) - u_N(t)\| \to 0, \qquad N \to \infty,$$

for $0 \leqq t \leqq T$. Thus, algebraic stability implies convergence in that subspace of $\mathcal{H}$ satisfying the conditions (5.3). If this latter subspace is large enough, an algebraically stable method can still be very useful although it cannot yield convergent results for all initial conditions $u(0)$ and forces $f(t)$. Since spectral methods are normally infinite-order accurate, algebraic stability implies convergence for such spectral methods.

In the examples of algebraic stability given in §§ 7–8, we find $r \leqq \frac{1}{2}$, $s \leqq 0$, and $K(t) \leqq K$. In this case, algebraic stability implies convergence so long as (5.3) holds with $p \leqq \frac{1}{2}$. Thus, the approximation need not be infinite-order accurate to achieve convergence. However, we develop the general theory of algebraic stability here in the expectation that it will find application to spectral methods for high-order equations in which p may be large.

Our definition of algebraic stability is very similar to the notion of s-stability introduced by Strang (1960). However, our motivation is slightly different. Strang introduced s-stability to study the convergence of time-discretized initial-value problems in which the norm of the evolution operator grows as a power of the time step.

Let us give an illustration of the need for a theory of algebraic stability. In §§ 8 and 11, we will study Chebyshev polynomial spectral methods to solve the one-dimensional wave equation $u_t + u_x = f(x, t)$ with boundary conditions $u(-1, t) = 0$. Unfortunately, this wave equation is *not* well posed in the Chebyshev norm

$$\|u\|^2 = \int_{-1}^{1} \frac{u^2(x)}{\sqrt{1-x^2}}\, dx.$$

In fact, if

$$u(x, 0) = \begin{cases} 1 - \dfrac{|x|}{\varepsilon} & \text{if } |x| < \varepsilon, \\ 0 & \text{if } |x| \geqq \varepsilon, \end{cases}$$

then the solution of $u_t + u_x = 0$, $u(-1, t) = 0$ at $t = 1$ is given by

$$u(x, 1) = \begin{cases} 1 - \dfrac{1}{\varepsilon} + \dfrac{x}{\varepsilon}, & 1 - \varepsilon < x \leqq 1, \\ 0, & x \leqq 1 - \varepsilon. \end{cases}$$

Therefore, as $\varepsilon \to 0+$,

$$\|u(x, 0)\|^2 \sim \varepsilon, \qquad \varepsilon \to 0+,$$

$$\|u(x, 1)\|^2 \sim \frac{2}{3}\sqrt{2\varepsilon}, \qquad \varepsilon \to 0+,$$

so that if $L = -\partial/\partial x$,

$$\|e^L\| \geqq \frac{\|u(x, 1)\|}{\|u(x, 0)\|} \sim \left(\frac{8}{9}\right)^{1/4} \varepsilon^{-1/4}, \qquad \varepsilon \to 0+. \tag{5.4}$$

In fact, $\|e^{Lt}\| = \infty$ for $0 < t < 2$, $\|e^{Lt}\| = 0$ for $t > 2$, so the one-dimensional wave equation is not well posed in the Chebyshev norm.

Since the finite-dimensional approximations L_N to L given by Galerkin, tau, and collocation approximation (see § 2) should converge as $N \to \infty$, it follows that we may expect

$$\|\exp(L_N t)\| \to \infty$$

as $N \to \infty$ in the Chebyshev norm. To estimate the rate of divergence of $\|\exp(L_N t)\|$

as $N \to \infty$ we argue that Chebyshev polynomials of degree at most N can resolve distances of at most order $1/N$ *interior* to $(-1, 1)$ so we may reasonably guess on the basis of (5.4) with $\varepsilon = 1/N$ that

$$\|\exp(L_N t)\| \geqq O(N^{1/4}), \qquad N \to \infty. \tag{5.5}$$

This result is justified by the numerical results presented in Table 8.3. Equation (5.5) implies that Chebyshev-spectral approximations to the one-dimensional wave equation are not stable but are algebraically stable with $r \geqq \frac{1}{4}$ and $s \geqq 0$ in (5.2).

Notice that algebraic stability in one norm implies algebraic stability in all algebraically equivalent norms. Thus, algebraic stability is equivalent in all of the L_p norms $1 \leqq p \leqq \infty$ because these norms are algebraically equivalent in N-dimensional vector spaces (i.e., they differ from each other only by a fixed power of N). To show this, we recall that the L_p norm of a vector $\mathbf{a} = (a_1, \cdots, a_N)$ is defined by

$$\|a\|_p = \left(\sum_{i=1}^{N} |a_i|^p \right)^{1/p}.$$

If $q = p\alpha$ with $0 < \alpha < 1$, then

$$\|a\|_q^q = \left(\sum_{i=1}^{N} |a_i|^{p\alpha} \right) \leqq \left(\sum_{i=1}^{N} |a_i|^p \right)^{\alpha} \left(\sum_{i=1}^{N} 1 \right)^{1-\alpha} = \|a\|_p^q N^{1-q/p}$$

by Hölder's inequality. Therefore, choosing $q = 1$, we have

$$N^{1/p-1}\|a\|_1 \leqq \|a\|_p, \qquad p \geqq 1.$$

Also, if $p \geqq 1$, then

$$\|a\|_p^p = \sum_{i=1}^{N} |a_i|^p \leqq \left(\sum_{i=1}^{N} |a_i| \right)^p = \|a\|_1^p,$$

so that

$$N^{1/p-1}\|a\|_1 \leqq \|a\|_p \leqq \|a\|_1, \qquad p \geqq 1. \tag{5.6}$$

The verification of algebraic stability for spectral methods is equivalent to a general problem in matrix theory. Suppose that A_N $(N = 1, 2, \cdots)$ is a one parameter family of matrices. The problem is to find conditions on the members of the family such that $\exp(A_N t)$ is algebraically stable. We will use only the L_2 norm since other matrix norms are algebraically equivalent to it.

Conditions for algebraic stability. Let $\{A_N\}$ be a family of $N \times N$ matrices where $\|A_N\| = O(N^a)$ $(N \to \infty)$ for some finite a. A necessary and sufficient condition for algebraic stability

$$\|e^{A_N t}\| = O(N^r N^{st}), \qquad N \to \infty,$$

is that there exist a family $\{H_N\}$ of Hermitian positive-definite matrices such that

(5.7a) $$\|H_N^{-1}\|\|H_N\| = O(N^b), \qquad N\to\infty,$$

(5.7b) $$H_N A_N + A_N^* H_N \leqq c(N) H_N,$$

(5.7c) $$c(N) < d \log N$$

for all sufficiently large N where b and d are finite numbers independent of N.

To prove the sufficiency of (5.7) we use the Lie formula

(5.8) $$e^{(C+D)t} = \lim_{n\to\infty} (e^{Ct/n}\, e^{Dt/n})^n,$$

which is valid for arbitrary matrices C and D. This formula is proved at the end of this section. We define

(5.9) $$\begin{aligned} C &= \tfrac{1}{2}[H_N^{1/2} A_N H_N^{-1/2} + H_N^{-1/2} A_N^* H_N^{1/2}],\\ D &= \tfrac{1}{2}[H_N^{1/2} A_N H_N^{-1/2} - H_N^{-1/2} A_N^* H_N^{1/2}]. \end{aligned}$$

Since

$$\exp[A_N t] = H_N^{-1/2} \exp[H_N^{1/2} A_N H_N^{-1/2}\, t] H_N^{1/2},$$

it follows from the Lie formula that

(5.10) $$e^{A_N t} = \lim_{n\to\infty} H_N^{-1/2} (e^{Ct/n}\, e^{Dt/n})^n H_N^{1/2}.$$

Since C is a symmetric matrix, it follows from (5.7b) that

$$\|e^{Ct/n}\| \leqq e^{ct/n}.$$

Also, D is an antisymmetric matrix so that

$$\|e^{Dt/n}\| = 1.$$

Therefore, (5.10) gives

(5.11) $$\|e^{A_N t}\| \leqq e^{ct} \|H_N^{-1/2}\|\|H_N^{1/2}\|$$

proving algebraic stability.

In order to prove that the conditions (5.7) are also necessary for algebraic stability we define

$$B_N = A_N - (s+1)\log(N) I.$$

Since $\|\exp(A_N(t)\| = O(N^{r+st})$,

$$\|e^{B_N t}\| = O\left(\frac{N^r}{N^t}\right), \qquad N\to\infty.$$

Since the eigenvalues of B_N have negative real parts, Lyapunov's theorem (Barnett and Storey (1974)) implies that there exists a Hermitian positive-definite matrix H_N such that

$$H_N B_N + B_N^* H_N = -I. \tag{5.12}$$

Thus,

$$H_N A_N + A_N^* H_N = -I + 2(s+1)\log(N) H_N \leqq c(N) H_N,$$

where $c(N) = 2(s+1)\log N$.

To complete the proof of (5.7) we need to estimate the norms of H_N and H_N^{-1}. An explicit formula for H_N is

$$H_N = \int_0^\infty e^{B_N^* t} e^{B_N t}\, dt; \tag{5.13}$$

in fact,

$$H_N B_N + B_N^* H_N = \int_0^\infty \frac{d}{dt}[e^{B_N^* t} e^{B_N t}]\, dt = -I.$$

Therefore,

$$\|H_N\| \leqq \int_0^\infty \|e^{B_N^* t}\| \|e^{B_N t}\|\, dt \leqq N^{2r} \int_0^\infty N^{-2t}\, dt \leqq N^{2r}$$

if $2 \ln N > 1$, i.e., $N \geqq 2$. Also, from (5.12) we obtain

$$B_N H_N^{-1} + H_N^{-1} B_N^* = -(H_N^{-1})^2,$$

so that

$$\|H_N^{-1}\|^2 \leqq 2\|B_N\| \|H_N^{-1}\|$$

or

$$\|H_N^{-1}\| \leqq 2\|B_N\| = O(N^a), \qquad N \to \infty. \tag{5.14}$$

This completes the proof of the necessity of (5.7).

The condition for algebraic stability given in (5.7) implies that for every algebraically stable problem, there is a new norm induced by the Lyapunov matrices H_N which is algebraically equivalent to the original norm and in which the problem is stable in the classical sense.

The above result gives a method for checking numerically the algebraic stability of a family $\{A_N\}$ of matrices satisfying $\|A_N\| = O(N^a)$ as $N \to \infty$:

(i) We check that the real parts of the eigenvalues of A_N are bounded from above by $s \log N$; otherwise, the family of matrices A_N are algebraically unstable.

(ii) We introduce $B_N = A_N - (s+1)\log(N) I$ and compute the Lyapunov matrix H_N such that $H_N B_N + B_N^* H_N = -I$. There are several numerically efficient techniques to compute H_N (Bartels and Stewart (1972)).

(iii) To verify algebraic stability, the condition number of H_N must be bounded by N^b for some finite b as $N \to \infty$. Noting (5.14), it is only necessary to verify that the eigenvalues of H_N are bounded from above by some finite power of N as $N \to \infty$.

This procedure is applied in §§ 7–8 to verify algebraic stability of model problems. Since (5.7) gives a necessary and sufficient condition for algebraic stability, if these conditions do not hold the family of matrices A_N is algebraically unstable.

Proof of the Lie formula. To prove the Lie formula (5.8) for finite-dimensional matrices, we use the identity

$$e^{C+D} - [e^{C/n} e^{D/n}]^n = [e^{((C+D)/n)}]^n - [e^{C/n} e^{D/n}]^n$$
$$= \sum_{k=0}^{n-1} [e^{((C+D)/n)}]^k (e^{(C+D)/n} - e^{C/n} e^{D/n})[e^{C/n} e^{D/n}]^{n-1-k}.$$

Thus,

$$\|e^{C+D} - (e^{C/n} e^{D/n})^n\| \leqq \sum_{k=0}^{n-1} e^{\|C+D\|k/n} \|e^{(C+D)/n} - e^{C/n} e^{D/n}\| (e^{\|C\|+\|D\|})^{(n-1-k)/n}$$
$$\leqq n\|e^{(C+D)/n} - e^{C/n} e^{D/n}\| \exp [(\|C\| + \|D\|)(1-1/n)].$$

On the other hand,

$$\|e^{(C+D)/n} - e^{C/n} e^{D/n}\| \leqq \frac{\|CD-DC\|}{2n^2} + O\left(\frac{1}{n^3}\right), \qquad n \to \infty,$$

so that

$$\|e^{C+D} - (e^{C/n} e^{D/n})^n\| \leqq K/n, \qquad n \to \infty,$$

for any $K > \frac{1}{2}\|CD - DC\| e^{\|C\|+\|D\|}$, proving (5.8).

Equation (5.8) is also true for certain infinite-dimensional matrices (operators). This deep result known as the Trotter product formula is very useful in the modern theory of partial differential equations.

SECTION 6

Spectral Methods Using Fourier Series

Fourier series are appropriate to solve problems with periodic boundary conditions. With periodic boundary conditions, a stable spectral method based on Fourier series is usually accurate and efficient. On the other hand, when Fourier series are used to solve nonperiodic problems (including problems having periodic initial conditions but whose evolution operators violate periodicity), stability is not enough to ensure convergence to the true solution of the problem. An example of the latter effect was given in Example 1.3. In this section, we investigate the stability and convergence of spectral methods based on Fourier series.

EXAMPLE 6.1. *Constant-coefficient hyperbolic equation with periodic boundary conditions.* Consider the one-dimensional wave equation

$$u_t + u_x = 0, \qquad 0 \leqq x \leqq 1, \tag{6.1}$$
$$u(x, 0) = f(x),$$

with periodic boundary conditions

$$u(0, t) = u(1, t).$$

Since collocation, Galerkin and tau methods are identical in the absence of essential boundary conditions (see § 2), let us analyze the Fourier-collocation or pseudospectral method.

We introduce the collocation points $x_n = n/2N$ $(n = 0, \cdots, 2N-1)$ and the vector notation $\mathbf{u} = (u_0, \cdots, u_{2N-1})$ where $u_n = u(x_n)$. The collocation equations that approximate (6.1) can be written as

$$\frac{\partial \mathbf{u}}{\partial t} = C^{-1} D C \mathbf{u}, \tag{6.2}$$

where C and D are $2N \times 2N$ matrices whose entries are

$$C_{kl} = \frac{1}{\sqrt{2N}} \exp\,[-2\pi i (k-N) x_l], \tag{6.3a}$$

$$D_{kl} = -2\pi i k' \delta_{kl}, \tag{6.3b}$$

where $k' = k - N$ $(1 \leqq k \leqq 2N-1)$ and $k' = 0$ if $k = 0$. A simple derivation of (6.2) is obtained by observing that $C\mathbf{u}$ gives the Fourier coefficients of the collocation projection Pu of $u(x)$. Thus, $DC\mathbf{u}$ are the Fourier coefficients of $-(\partial/\partial x)Pu$ and, finally, $C^{-1}\ DC\mathbf{u}$ gives the collocation projection of $-(\partial/\partial x)Pu$ which is $-P(\partial/\partial x)Pu$. The matrix C is a unitary matrix so $C^* = C^{-1}$, and the matrix D is skew-Hermitian so $D^* = -D$. Therefore, $C^{-1}DC$ is skew-Hermitian so that

$$\|\exp[C^{-1}DC]t\| = 1. \tag{6.4}$$

This proves that the Fourier-collocation method is stable for (6.1). The results of this example can be generalized to a general system of constant coefficient hyperbolic equations.

EXAMPLE 6.2. *Variable-coefficient hyperbolic equation with periodic boundary conditions.* Consider the system of equations

$$u_t = A(x)u_x, \qquad 0 \leqq x \leqq 1,$$

with periodic boundary conditions $u(0, t) = u(1, t)$ and periodic inhomogenity: $A(x) = A(x+1)$ for all x. Here $u(x)$ is a vector of m components and $A(x)$ is an $m \times m$ matrix. If we assume that $A(x)$ is a symmetric matrix and that

$$\frac{\partial A}{\partial x} \leqq \alpha I \tag{6.5}$$

for some finite α, then the Fourier–Galerkin method is stable. To show this, we denote by u_N the N-term Fourier–Galerkin approximation of u. Using (2.7)–(2.8) and integration by parts, we obtain

$$\frac{d}{dt}\int_0^1 u_N^* u_N \, dx = \int_0^1 u_N^*(A + A^*)_x u_N \, dx \leqq 2\alpha \int_0^1 u_N^* u_N \, dx.$$

Therefore,

$$\int_0^1 u_N^*(x, t)u_N(x, t) \, dx \leqq e^{2\alpha t}\int_0^1 u_N^*(x, 0)u_N(x, 0) \, dx$$

which proves stability.

Condition (6.5) is not sufficient to ensure stability for the collocation method. Consider the scalar equation ($m = 1$)

$$\begin{aligned} u_t + r(x)u_x &= 0, \qquad 0 \leqq x \leqq 1, \\ u(0, t) &= u(1, t). \end{aligned} \tag{6.6}$$

If we impose the additional restriction that $r(x)$ is nonzero within $0 \leqq x \leqq 1$, then we can prove that the collocation method is stable. To do this, we show that $\exp(RC^*DCt)$ is stable where C and D are given by (6.3) and R is the matrix with entries

$$R_{ij} = r(x_i)\delta_{ij}.$$

The matrix R^{-1} can be identified as the Lyapunov matrix H_N invoked in (5.7) and, therefore, the method is stable:

$$R^{-1}(RC^*DC) + (C^*D^*CR^*)R^{-1} = 0.$$

In fact, following the argument leading to (5.11) gives

$$\|\exp(RC^*DCt)\|^2 \leqq \|R\|\|R^{-1}\| \leqq \max_{0\leqq x\leqq 1} |r(x)|/\min_{0\leqq x\leqq 1} |r(x)|,$$

proving stability for $N\to\infty$.

If $r(x)$ has a zero within $0<x<1$, collocation with Fourier series may lead to instability. For example, if $N=2$, the eigenvalues of RC^*DC are $0, 0, \pm\sqrt{-(r_0+r_2)(r_1+r_3)}$ where $r_i=r(x_i)$, so there are growing modes if $(r_0+r_2)(r_1+r_3)<0$. In some cases, these modes may have large growth rates. One way to limit the growth rate of these modes is to rewrite (6.6) as

$$u_t+\frac{1}{2}(r(x)u)_x+\frac{1}{2}r(x)u_x-\frac{1}{2}\frac{dr(x)}{dx}u=0. \tag{6.7}$$

Now Fourier-collocation gives the matrix equation

$$\mathbf{u}_t=(\tfrac{1}{2}C^*DCR+\tfrac{1}{2}RC^*DC+Q)\mathbf{u},$$

where $Q_{kl}=\frac{1}{2}r'(x_k)\delta_{kl}$. The first two matrices on the right side add up to a skew-Hermitian matrix. Also, if (6.5) holds for $r(x)$ then $Q\leqq\frac{1}{2}\alpha I$. Therefore, we obtain the inequality

$$\frac{d}{dt}|\mathbf{u}|^2\leqq\alpha|\mathbf{u}|^2.$$

Thus, we see it is possible to bound a priori the growth of modes in the Fourier-collocation method for variable coefficient problems with periodic boundary conditions.

On the other hand, for problems with nonperiodic boundary conditions, Fourier-spectral methods can produce wrong solutions even when they are stable. This is illustrated by Example 1.3 which we now study more carefully.

EXAMPLE 6.3. *Hyperbolic equation with nonperiodic boundary conditions.* Consider the problem (1.8):

$$\begin{aligned} &\frac{\partial u(x,t)}{\partial t}+\frac{\partial u(x,t)}{\partial x}=x+t, && 0<x<\pi,\quad t>0,\\ &u(0,t)=0, && t\geqq 0,\\ &u(x,0)=0, && 0\leqq x\leqq\pi. \end{aligned} \tag{6.8}$$

The solution is

$$u(x,t)=xt.$$

If we attempt to solve (6.8) by Fourier sine series using the Galerkin procedure we obtain

$$u_N = \sum_{n=1}^{N} a_n \sin nx, \tag{6.9}$$

$$\frac{da_n}{dt} = -\frac{4}{\pi} \sum_{\substack{m=1 \\ m+n \text{ odd}}}^{N} \frac{nm}{n^2-m^2} a_m - \frac{2}{n}(-1)^n + \frac{4}{\pi n} t e_n, \tag{6.10}$$

where $e_n = 0$ if n is even and $e_n = 1$ if n is odd.

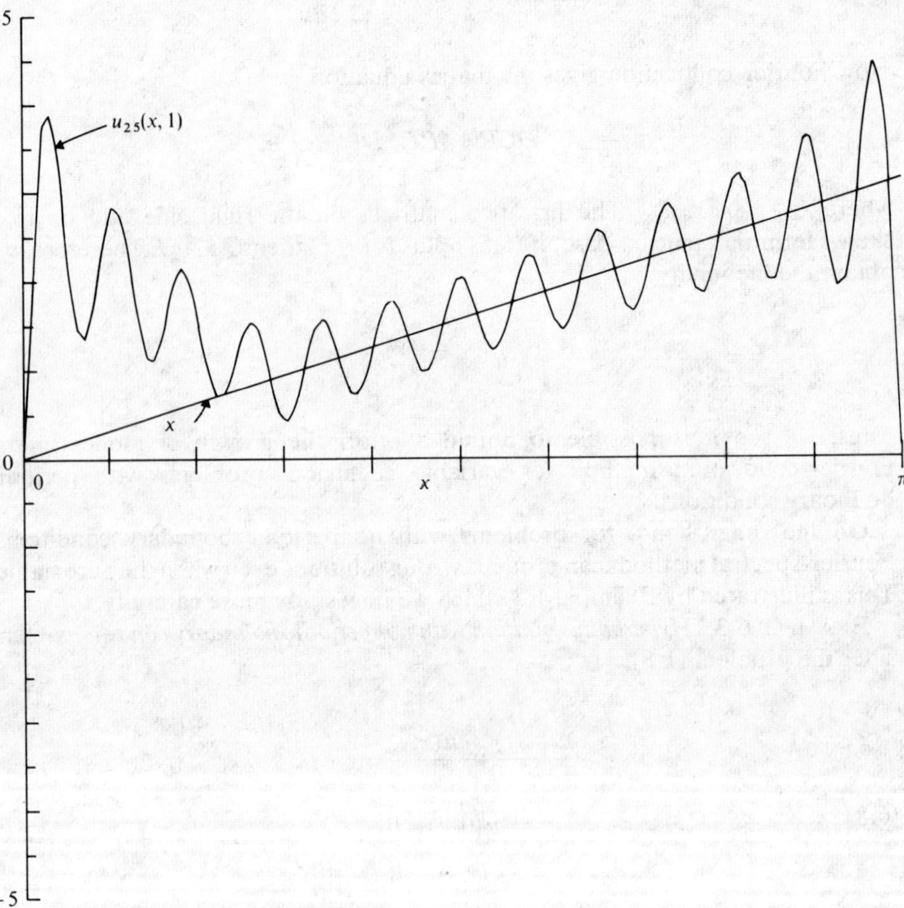

FIG. 6.1. *A plot of $u_N(x, t)$ vs. x for $N = 25$ and $t = 1$ where $u_N(x, t)$ is determined by numerical integration of (6.9)–(6.10) with negligible time-differencing errors. A plot of the exact solution xt at $t = 1$ to (6.8) is also given. Observe the apparent divergence of $u_N(x, t)$ from the exact solution for $0 \leqq x < t$ and the enhanced Gibbs phenomenon at $x = 0, \pi$.*

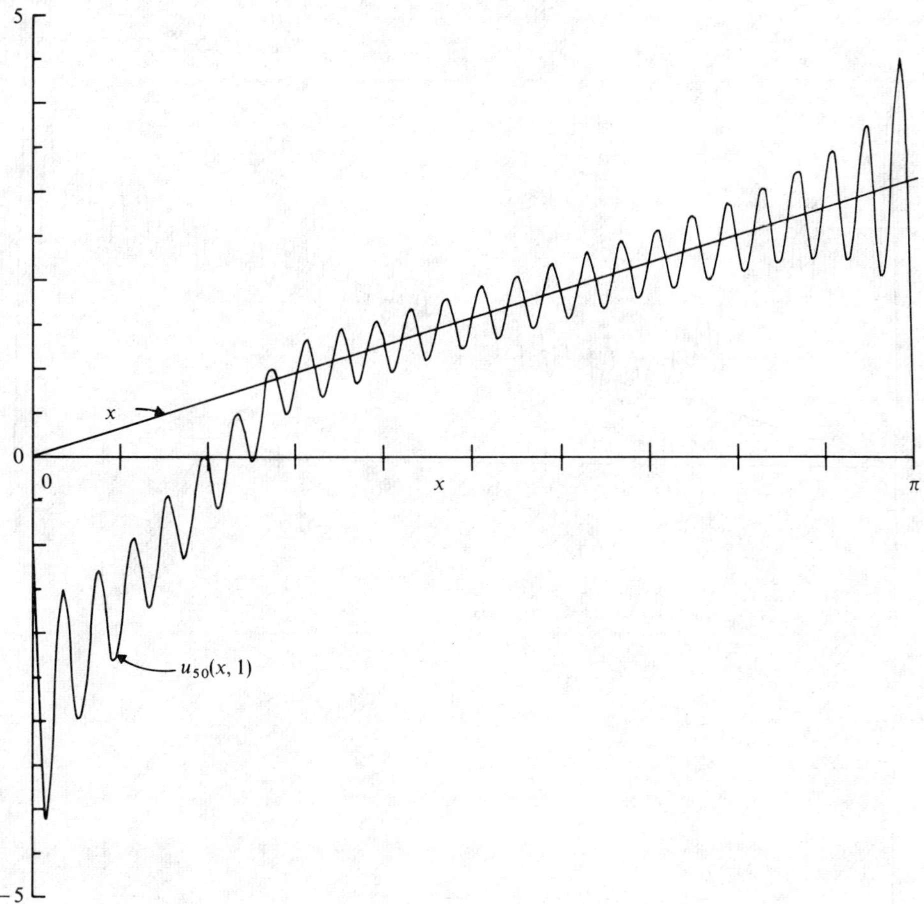

FIG. 6.2. *Same as Fig.* 6.1 *except* $N = 50$, $t = 1$.

It is easy to verify that the above approximation is stable. If we write (6.10) in the form

$$\frac{d\mathbf{a}}{dt} = A_N\mathbf{a} + \mathbf{f}$$

where $\mathbf{a} = (a_1, \cdots, a_N)$, $\mathbf{f} = (f_1, \cdots, f_N)$, $f_n = (2/n)[(-1)^{n+1} + 2te_n/\pi]$, then

$$A_N + A_N^* = 0.$$

Thus, $\|\exp(A_N t)\| = 1$ for all N and t.

In Figs. 6.1–6.4 we plot the solution of (6.9)–(6.10) at $t = 1$ for $N = 25, 50, 75, 100$. It is apparent that $u_N(x, 1)$ does not converge to the exact solution xt at $t = 1$

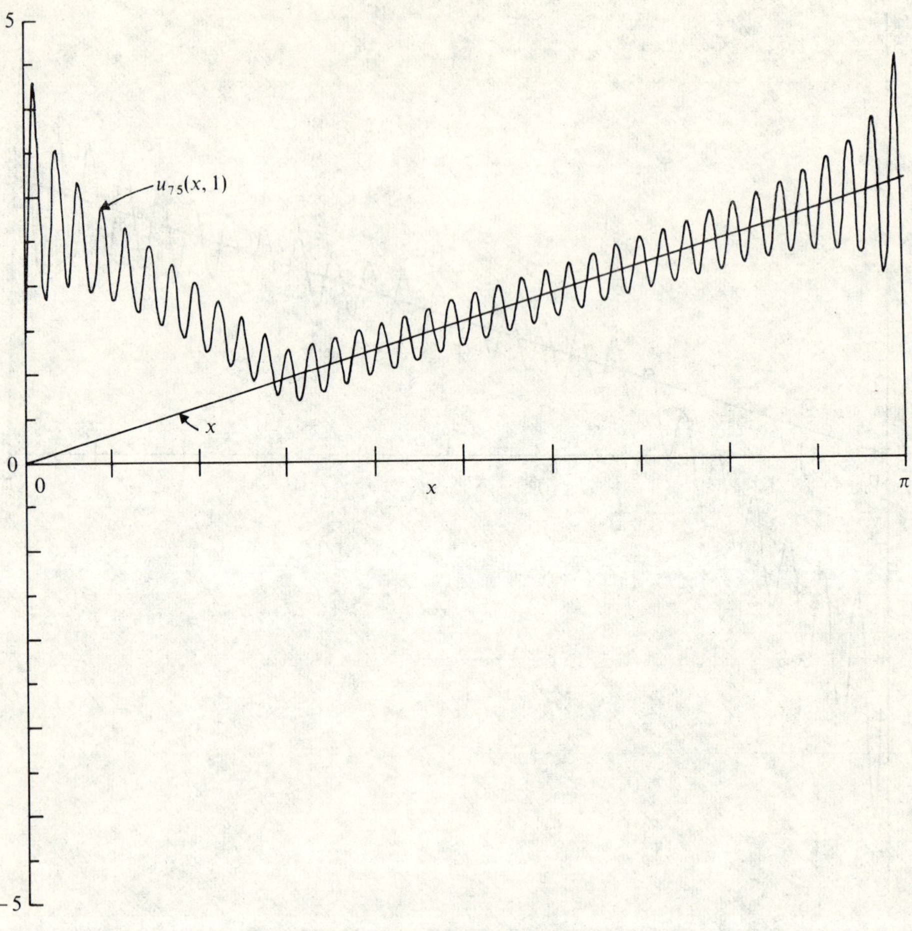

FIG. 6.3. *Same as Fig.* 6.1 *except* $N = 75$, $t = 1$.

as $N \to \infty$. Instead, u_N for N even appears to be converging as $N \to \infty$ to the function

$$u_{\text{even}} = \begin{cases} xt, & x \geqq t, \\ \pi(x-t)+xt, & x < t, \end{cases} \tag{6.11}$$

for $t < \pi$, while u_N for N odd appears to converge to the function

$$u_{\text{odd}} = \begin{cases} xt, & x \geqq t, \\ \pi(t-x)+xt, & x < t, \end{cases} \tag{6.12}$$

for $t < \pi$. The results plotted in Fig. 6.5 for $u_{100}(x, t=2)$ are also consistent with convergence to the wrong solution (6.11). Notice that the approximations $u_N(x, t)$ plotted in Figs. 6.1–5 all exhibit a large region of nonuniform convergence near $x = 0$ and $x = \pi$ and that the rate of convergence to the wrong solutions (6.11)–(6.12) in the interior of the interval $0 < x < \pi$ is roughly like $1/\sqrt{N}$.

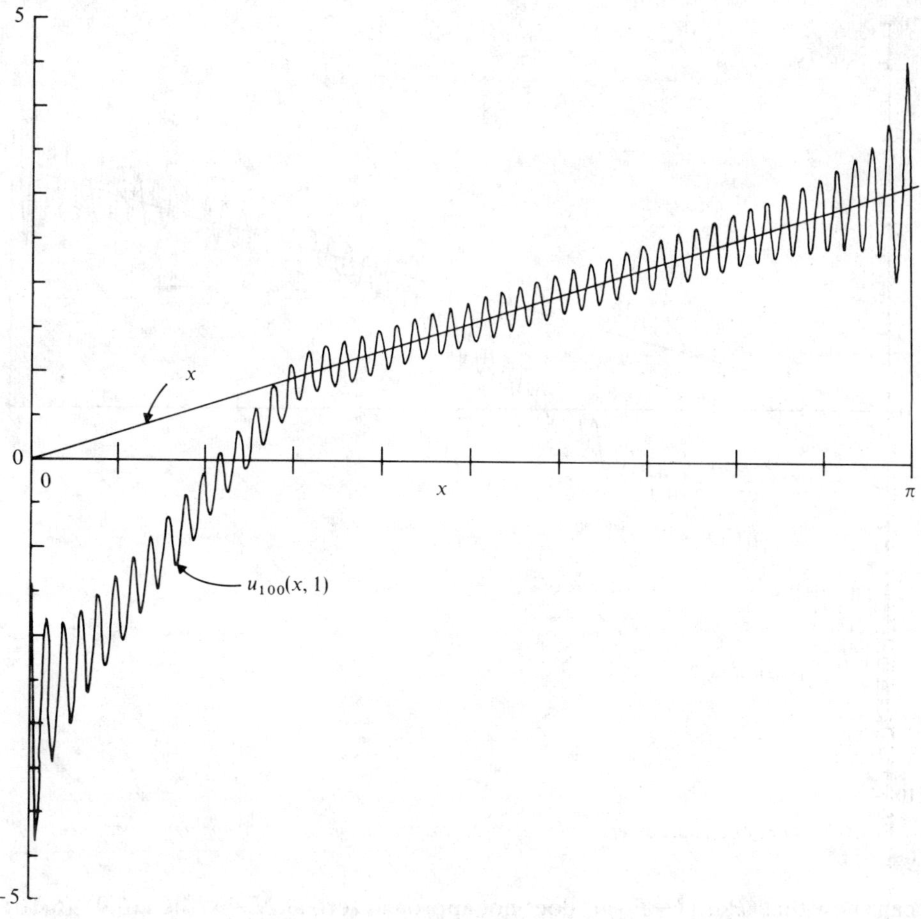

FIG. 6.4. *Same as Fig.* 6.1 *except* $N = 100$, $t = 1$.

The origin of the divergence of (6.9)–(6.10) from the exact solution to (6.8) is not instability; rather, the divergence is due to inconsistency. Since $\|\exp(A_N t)\| = 1$, the method is stable. To show that it is not consistent we estimate the truncation error in the L_2 norm,

$$\varepsilon_N = \|Lu - L_N u\|,$$

for $u = xt$ where $L_N = P_N L P_N$ and P_N is the Galerkin projection operator and $L = -\partial/\partial x$. This error can be bounded from below by

$$\begin{aligned}\varepsilon_N &= \|Lu - P_N Lu + P_N Lu - P_N L P_N u\| \\ &\geq \|P_N L(I - P_N)u\| - \|(I - P_N)Lu\|.\end{aligned}$$

However, $\|(I - P_N)Lu\| \to 0$ (like $1/\sqrt{N}$) as $N \to \infty$ because this norm is just the error in the Fourier sine series expansion of $Lu = -(\partial/\partial x)xt = t$. Therefore, if we

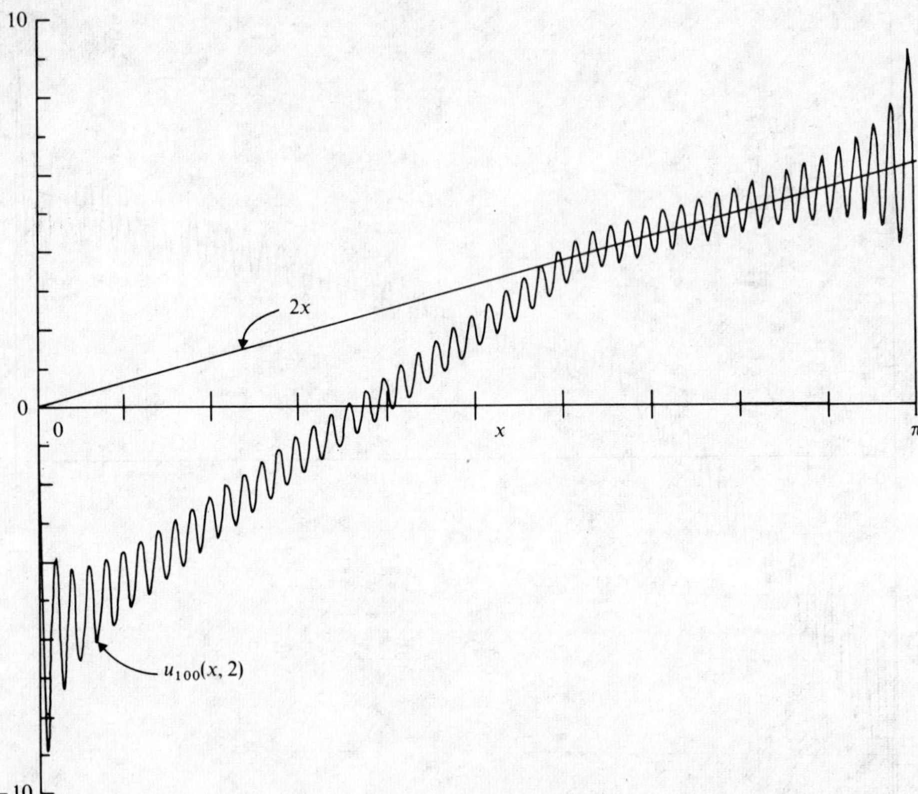

FIG. 6.5. *Same as Fig. 6.1 except $N = 100$, $t = 2$. Observe that the region of apparent divergence is still $0 \leqq x < t$.*

can show that $\|P_N L(I - P_N)u\|$ does not approach zero as $N \to \infty$ then (6.9)–(6.10) is not consistent.

To estimate $\|P_N L(I - P_N)u\|$ we proceed as follows. Since

$$(I - P_N)u = \sum_{n=N+1}^{\infty} a_n(t) \sin nx$$

we obtain

$$P_N L(I - P_N)u = \sum_{n=1}^{N} b_n(t) \sin nx,$$

where

$$b_n(t) = -\frac{4}{\pi} \sum_{\substack{m=N+1 \\ m+n \text{ odd}}}^{\infty} \frac{nm}{n^2 - m^2} a_m(t).$$

Therefore, since the Fourier coefficients of u are given by $a_n(t)=2(-1)^{n+1}t/n$,

$$\|P_N(I-P_N)u\|^2=\sum_{n=1}^{N} b_n^2$$

$$=\frac{64}{\pi^2}t^2\sum_{n=1}^{N}\left(\sum_{\substack{m=N+1\\ m+n\ \text{odd}}}^{\infty}\frac{n}{n^2-m^2}\right)^2$$

$$\geqq\frac{64t^2}{\pi^2}\sum_{n=1}^{N}\left(\sum_{\substack{m=N+1\\ m+n\ \text{odd}}}^{\infty}\frac{n}{m^2}\right)^2$$

$$\geqq Ct^2\sum_{n=1}^{N}\frac{n^2}{N^2}\geqq C_1t^2N$$

for suitable constants C and C_1. This completes the proof that $\|Lu-L_Nu\|$ does not approach zero as $N\to\infty$.

Blair Swartz (private communication, 1976) traces the inconsistency of (6.9)–(6.10) to the incompleteness of the set of functions $\{L(\sin nx)=-n\cos nx, n=1, 2, \cdots\}$. This set of functions is made complete by augmenting it by the function 1. Whereas u may be well approximated by a function u_N of the form (6.9), Lu may not be well approximated by the function Lu_N. In fact, if $\|Lu - Lu_N\|\to 0$ as $N\to\infty$, then

$$\int_0^{\pi}(Lu-Lu_N)\,dx\to 0,\qquad N\to\infty.$$

Since

$$\int_0^{\pi}Lu_N=-\int_0^{\pi}\sum na_n\cos nx\,dx=0,$$

Lu may be well approximated by Lu_N only if

$$0=\int_0^{\pi}Lu\,dx=u(0)-u(\pi),$$

which is generally not true.

As shown in Figs. 6.1–5, $u_N(x,t)$ does not converge to xt as $N\to\infty$. The analysis given above provides no clue to the fascinating way in which the method achieves this divergence. There is no indication of the "error" wave $(-1)^N\pi(x-t)$ that appears in (6.11)–(6.12) and propagates with speed 1 across $0<x<\pi$. It seems that the complete mathematical analysis of the divergence of (6.9)–(6.10) is difficult and we do not now have a justifiable argument to demonstrate convergence of u_N to u_{even} and u_{odd} given by (6.11)–(6.12) as $N\to\infty$ through even and odd values, respectively.

In the next example we will show that it is not simply the presence of boundary conditions but rather the nonperiodic nature of the problem that causes the divergence of the Fourier-spectral methods.

EXAMPLE 6.4. *Nonperiodic boundary-free problem.* Consider the problem

$$\frac{\partial u}{\partial t}+\left(x-\frac{\pi}{2}\right)\frac{\partial u}{\partial x}=0, \qquad 0<x<\pi,$$

(6.13)

$$u(x,0)=f(x).$$

The problem is well posed without specifying any boundary condition. However, it is clear that the exact solution given by

(6.14) $$u(x,t)=f\left(\frac{\pi}{2}+\left(x-\frac{\pi}{2}\right)e^{-t}\right)$$

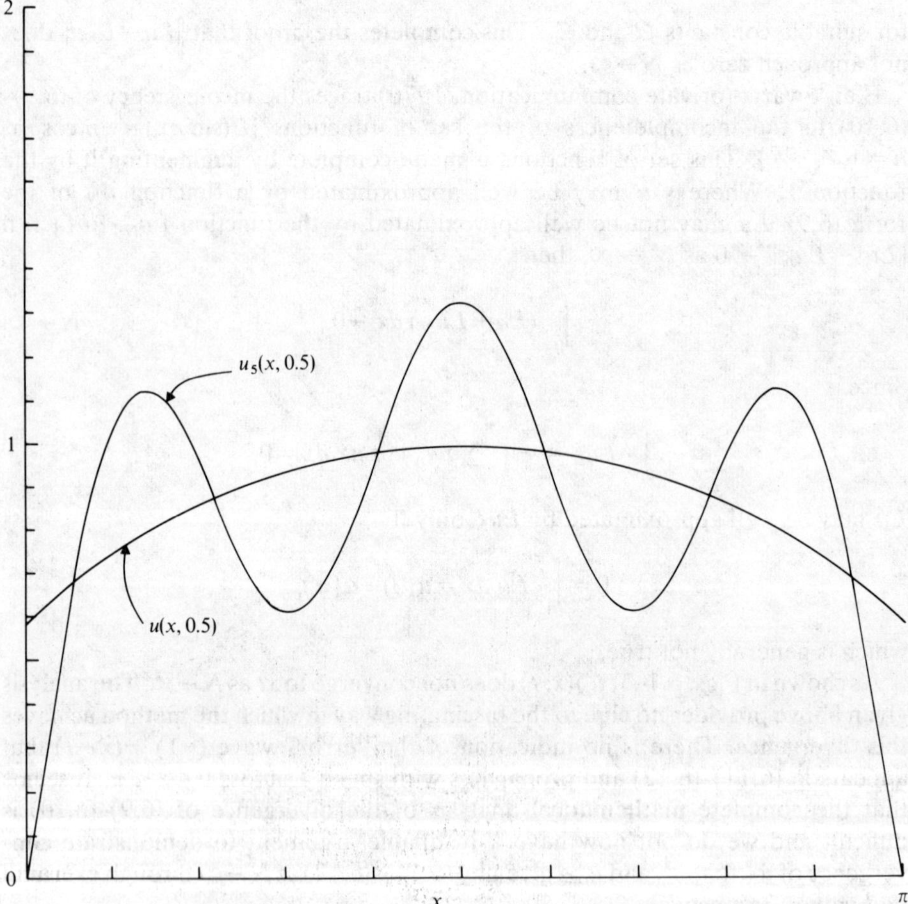

FIG. 6.6. *A plot of $u(x, t)$ and $u_N(x, t)$ vs. x for $N=5$, $t=0.5$. Here $u(x, t)$ is the exact solution of (6.13) and $u_N(x, t)$ is the Galerkin approximation to this solution using an N term Fourier sine series expansion. Observe the apparent divergence of $u_N(x, t)$ from $u(x, t)$.*

is not periodic in x. Since $r(x)=x-\pi/2$ has a bounded derivative, it follows from Example 6.2 that Fourier–Galerkin approximation to (6.13) is stable. Nevertheless it is *not* convergent as shown by the results plotted in Figs. 6.6–6.8 for $f(x)=\sin x$ and $N=5$, 10, and 20 retained terms in the Fourier sine series.

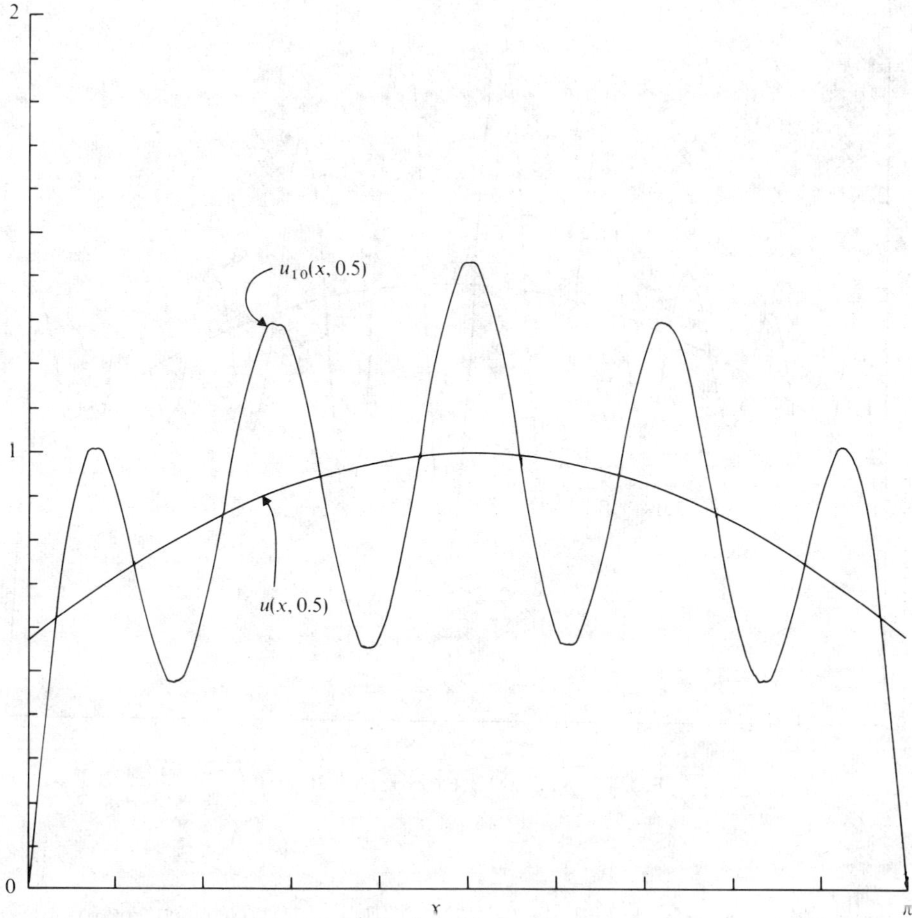

FIG. 6.7. *Same as Fig.* 6.6 *except* $N=10$, $t=0.5$.

Polynomial subtractions for nonperiodic problems. There is a method that can be used to ensure that Fourier series yield convergent results for nonperiodic problems. The idea is to express the solution as the sum of a low-order polynomial and a Fourier series; the polynomial is chosen so that the Fourier series converges rapidly as suggested originally by Lanczos (1956), (1966). The method has been used by Orszag (1971c) and Wengle and Seinfeld (1977) to solve problems with nonperiodic boundary conditions. We illustrate it here for the problem discussed in Example 6.4.

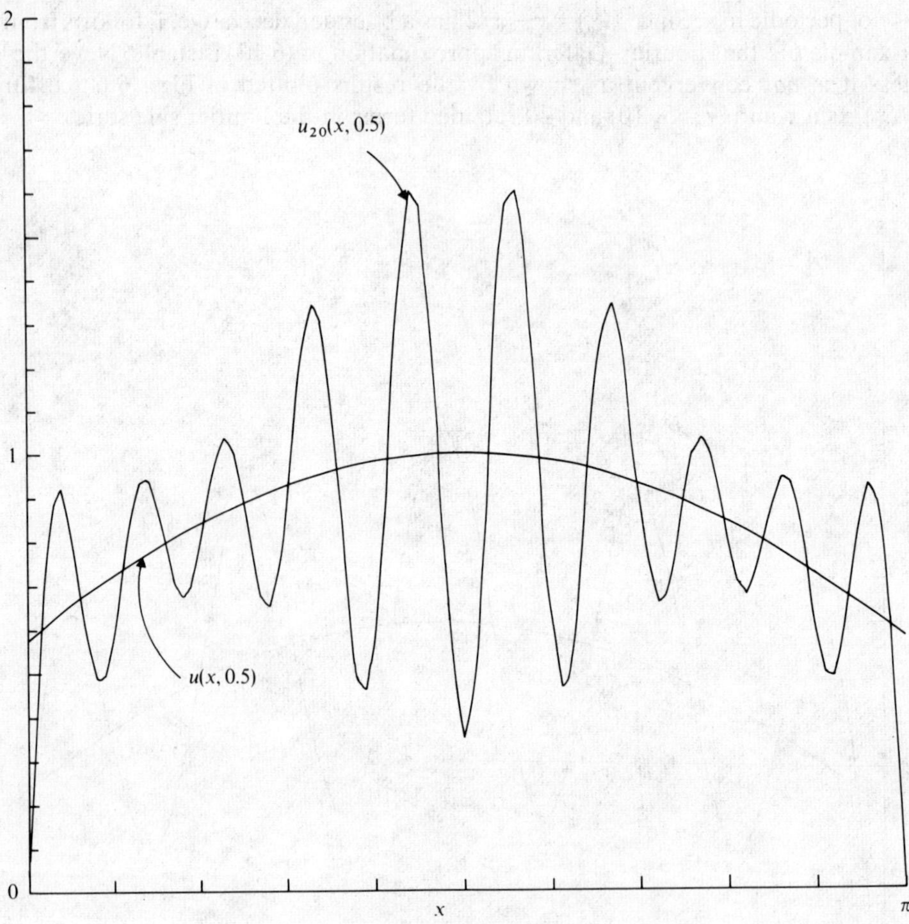

FIG. 6.8. *Same as Fig. 6.6 except $N=20$, $t=0.5$.*

EXAMPLE 6.5. *Polynomial subtractions applied to Fourier series.* The Fourier sine series expansion of the exact solution $u(x, t)$ to (6.13) converges slowly because, in general, $u(0, t)\neq 0$ and $u(\pi, t)\neq 0$. This slow convergence of the Fourier series of the exact solution implies that Galerkin approximation is inconsistent, as shown using the methods of Example 6.3. In order to avoid slow convergence or even divergence, we proceed as follows.

We seek the solution to (6.13) as the sum of a linear polynomial and a Fourier series:

$$u(x, t)=b(t)x+c(t)(\pi-x)+\sum_{n=1}^{\infty} a_n(t)\sin nx, \tag{6.15}$$

where $b(t)$ and $c(t)$ are chosen to ensure that $a_n(t)\to 0$ rapidly as $n\to\infty$. Substituting (6.15) into (6.13) gives

(6.16)
$$b'(t)x+c'(t)(\pi-x)+\sum_{n=1}^{\infty} a_n'(t)\sin nx=\left(\frac{\pi}{2}-x\right)[b(t)-c(t)] + \sum_{n=1}^{\infty} \hat{a}_n(t)\sin nx$$

where

(6.17)
$$\hat{a}_n(t)=\sum_{\substack{m=1\\ n+m \text{ even}\\ n\neq m}}^{\infty} \frac{2nm}{n^2-m^2}a_m+\frac{1}{2}a_n$$

are the Fourier sine coefficients of $(\pi/2-x)(\partial/\partial x)\sum a_n \sin nx$. If we knew $u(0,t)$ and $u(\pi,t)$ we could set $b(t)=u(\pi,t)/\pi$ and $c(t)=u(0,t)/\pi$; with this choice, the Fourier sine series in (6.15) does not exhibit the Gibbs phenomenon and $a_n(t)=O(1/n^3)$ as $n\to\infty$. However, the boundary conditions on u are not known as part of the specifications of the problem (6.13). Therefore, we must solve for $b(t)$ and $c(t)$ directly from the differential equation.

Equating coefficients of sin nx in (6.16) gives

(6.18)
$$\frac{da_n}{dt}=[c'-b'+c-b]\frac{2}{n}(-1)^{n+1}+[b-c-2c']\frac{2}{n}e_n+\hat{a}_n,\quad n=1,\cdots,$$

where $e_n=1$ if n is odd, 0 if n is even; here we use the Fourier sine series expansion of 1 and x:

$$1=\frac{4}{\pi}\sum_{\substack{n=1\\ n \text{ odd}}}^{\infty}\frac{\sin nx}{n},$$

$$x=2\sum_{n=1}^{\infty}(-1)^{n+1}\frac{\sin nx}{n}.$$

Also, if $b(t)$ and $c(t)$ are chosen so that $a_n=O(1/n^3)$ as $n\to\infty$, then the Fourier series $\sum a_n \sin nx$ may be differentiated termwise so

$$\sum_{n=1}^{\infty}\hat{a}_n\sin nx=\left(\frac{\pi}{2}-x\right)\frac{\partial}{\partial x}\sum_{n=1}^{\infty}a_n\sin nx=\left(\frac{\pi}{2}-x\right)\sum_{n=1}^{\infty}na_n\cos nx.$$

Therefore,

$$\lim_{x\to 0+}\sum_{n=1}^{\infty}\hat{a}_n\sin nx=\frac{\pi}{2}\sum_{n=1}^{\infty}na_n,$$

$$\lim_{x\to\pi-}\sum_{n=1}^{\infty}\hat{a}_n\sin nx=-\frac{\pi}{2}\sum_{n=1}^{\infty}(-1)^n na_n.$$

Using these results and setting $x=\pi$ and $x=0$ in (6.16) gives, respectively,

$$\frac{db}{dt}=\frac{1}{2}(c-b)-\frac{1}{2}\sum_{n=1}^{\infty}(-1)^n na_n, \tag{6.19}$$

$$\frac{dc}{dt}=\frac{1}{2}(b-c)+\frac{1}{2}\sum_{n=1}^{\infty} na_n. \tag{6.20}$$

Galerkin approximation reproduces the equations (6.17)–(6.20) with $a_n=0$ for $n=N+1, N+2, \cdots$.

The above derivation suggests, but does not prove, that $a_n(t)\to 0$ sufficiently rapidly as $n\to\infty$ so that inconsistency problems are avoided. The exact solution of (6.13), which satisfies (6.17)–(6.20) with $N=\infty$, does satisfy $a_n=O(1/n^3)$ as $n\to\infty$. However, the Galerkin approximation with finite N does not yield such a rapidly converging result. In fact, estimates like those given in Example 6.3 show that

$$\|Lv-L_N v\|=O\left(\frac{1}{N^{3/2}}\right), \qquad N\to\infty, \tag{6.21}$$

where v satisfies $v(0,t)=v(\pi,t)=0$ and $L=(\pi/2-x)\partial/\partial x$. Since the Galerkin approximation (6.18) is stable (see Example 6.6), we expect that the errors in the Galerkin approximation (6.17)–(6.20) are of order $N^{-3/2}$ for fixed t.

The above prediction has been tested numerically. In Table 6.1 we list for various N the maximum errors in the approximation obtained by solving (6.17)–(6.20). A plot of the error $u_N(x,t)-u(x,t)$ vs. x for $N=30, 40$ at $t=0.5$ is given in Figs. 6.9–10.

TABLE 6.1

Errors in the polynomial-subtracted Fourier series approximation $u_N(x,t)$ given by (6.22) and (6.17)–(6.20) for the problem (6.13) with $f(x)=\sin x$ for $t=0.5$. Observe that the errors appear to decrease as $N^{-3/2}$ as $N\to\infty$ in agreement with the estimate (6.21).

N	$\varepsilon_N=\max\lvert u_N(x,t=0.5)-u(x,t=0.5)\rvert$	$N^{3/2}\varepsilon_N$
5	4.19 (−3)	4.7 (−2)
10	2.13 (−3)	6.7 (−2)
15	1.13 (−3)	6.6 (−2)
20	8.28 (−4)	7.4 (−2)
25	5.76 (−4)	7.2 (−2)
30	4.70 (−4)	7.7 (−2)
35	3.64 (−4)	7.5 (−2)
40	3.13 (−4)	7.9 (−2)

In the next example, we prove that the method of polynomial subtraction used in Example 6.5 is stable.

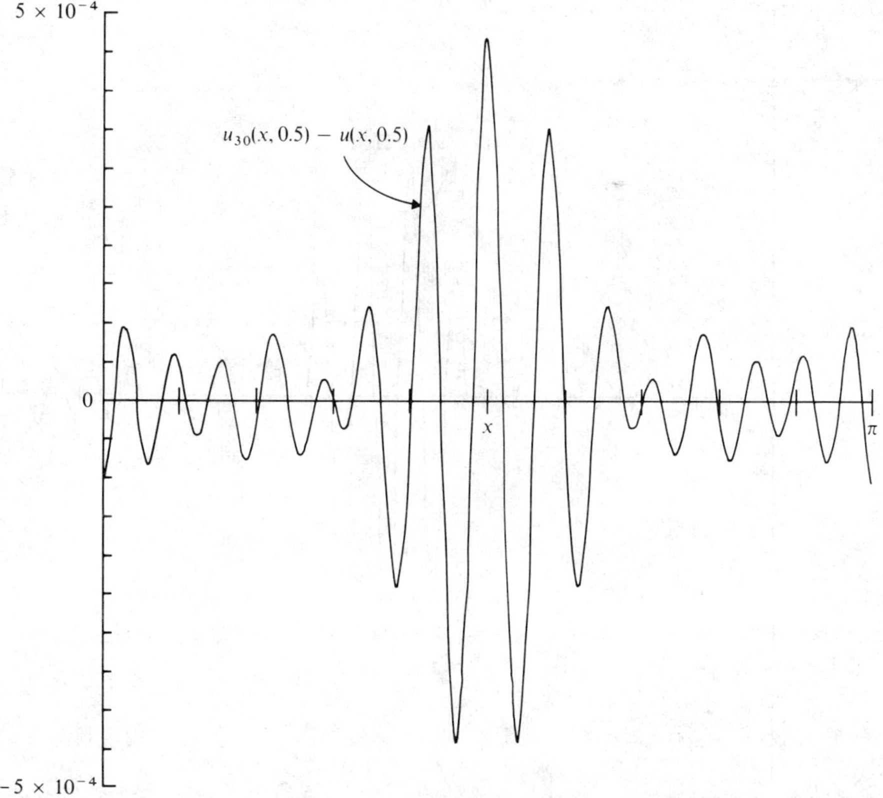

FIG. 6.9. *A plot of the error $u_N(x, t) - u(x, t)$ in the polynomial-subtracted Fourier-series approximation to* (6.13) *with $f(x) = \sin x$, $N = 30$ and $t = 0.5$. The approximation $u_N(x, t)$ is given by* (6.22) *and satisfies* (6.18)–(6.20); *the exact solution $u(x, t)$ is given by* (6.14).

EXAMPLE 6.6. *Proof of stability for polynomial subtractions.* It is not obvious that the approximation (6.17)–(6.20) is stable. Fourier series approximation without polynomial subtractions are stable but not consistent (see Example 6.4). On the other hand, the approximations obtained by polynomial subtractions are consistent as shown by (6.21), but their stability remains to be shown.

To demonstrate stability of (6.17)–(6.20), we reformulate these equations in terms of $u_N(x, t)$ defined by

$$u_N(x, t) = b(t)x + c(t)(\pi - x) + \sum_{n=1}^{N} a_n(t) \sin nx. \tag{6.22}$$

In terms of $u_N(x, t)$, (6.18) is equivalent to

$$\int_0^\pi \left[\frac{\partial u_N}{\partial t} + \left(x - \frac{\pi}{2}\right)\frac{\partial u_N}{\partial x}\right] \sin nx = 0, \qquad n = 1, \cdots, N, \tag{6.23}$$

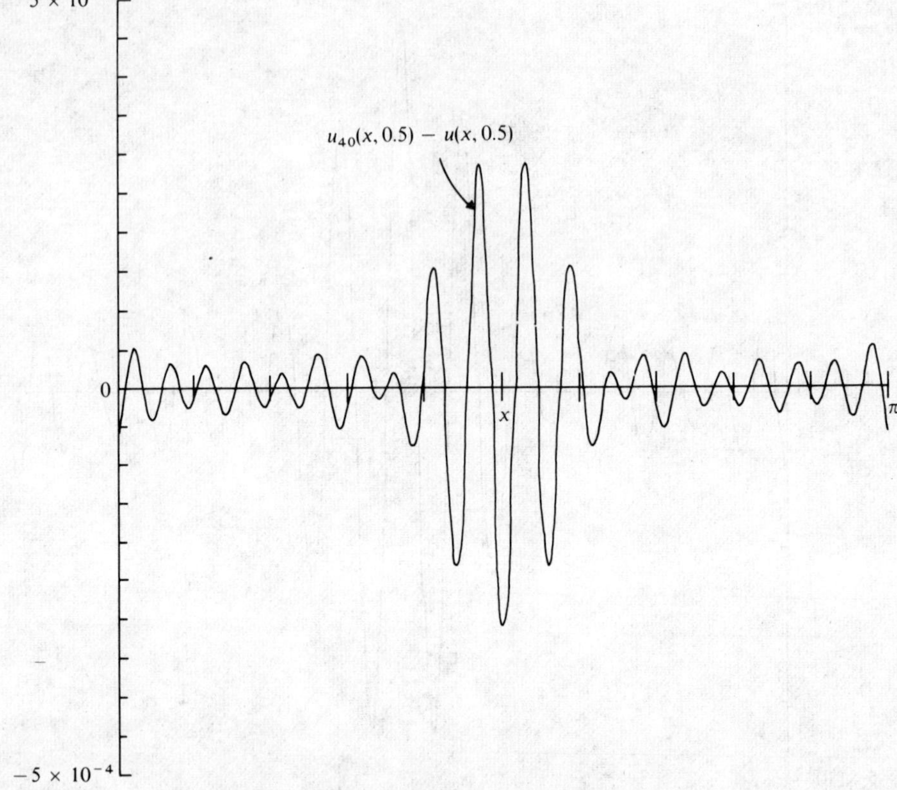

FIG. 6.10. *Same as Fig. 6.9 except* $N = 40$, $t = 0.5$.

while (6.19)–(6.20) become, respectively,

$$\left.\left[\frac{\partial u_N}{\partial t} + \left(x - \frac{\pi}{2}\right)\frac{\partial u_N}{\partial x}\right]\right|_{x=\pi} = 0, \tag{6.24}$$

$$\left.\left[\frac{\partial u_N}{\partial t} + \left(x - \frac{\pi}{2}\right)\frac{\partial u_N}{\partial x}\right]\right|_{x=0} = 0. \tag{6.25}$$

Multiplying (6.23) by $n^2 a_n$, summing from $n = 1$ to $n = N$, and noting that

$$\frac{\partial^2 u_N}{\partial x^2} = -\sum_{n=0}^{N} n^2 a_n \sin nx,$$

we obtain

$$\int_0^\pi \left[\frac{\partial u_N}{\partial t} + \left(x - \frac{\pi}{2}\right)\frac{\partial u_N}{\partial x}\right]\frac{\partial^2 u_N}{\partial x^2}\,dx = 0. \tag{6.26}$$

Integrating (6.26) once by parts and using (6.24)–(6.25), we obtain

$$\int_0^\pi \frac{\partial}{\partial x}\left[\frac{\partial u_N}{\partial t} + \left(x - \frac{\pi}{2}\right)\frac{\partial u_N}{\partial x}\right]\frac{\partial u_N}{\partial x}\,dx = 0.$$

Therefore,

$$\frac{\partial}{\partial t}\int_0^\pi \left(\frac{\partial u_N}{\partial x}\right)^2 dx = -2\int_0^\pi \left(\frac{\partial u_N}{\partial x}\right)^2 dx - \int_0^\pi \left(x-\frac{\pi}{2}\right)\frac{\partial}{\partial x}\left(\frac{\partial u_N}{\partial x}\right)^2 dx.$$

Integrating the second integral on the right once by parts gives

$$\frac{\partial}{\partial t}\int_0^\pi \left(\frac{\partial u_N}{\partial x}\right)^2 dx = -\int_0^\pi \left(\frac{\partial u_N}{\partial x}\right)^2 dx - \frac{\pi}{2}\left[\left(\frac{\partial u_N}{\partial x}\right)^2\bigg|_{x=\pi} + \left(\frac{\partial u_N}{\partial x}\right)^2\bigg|_{x=0}\right],$$

so that

$$\frac{\partial}{\partial t}\int_0^\pi \left(\frac{\partial u_N}{\partial x}\right)^2 dx \leqq -\int_0^\pi \left(\frac{\partial u_N}{\partial x}\right)^2 dx.$$

Thus, we obtain the stability estimate

$$\int_0^\pi \left[\frac{\partial u_N(x,t)}{\partial x}\right]^2 dx \leqq e^{-t}\int_0^\pi \left[\frac{\partial u_N(x,0)}{\partial x}\right]^2 dx. \tag{6.27}$$

The bound (6.27) shows the stability of (6.17)–(6.20).

Examples 6.5–6.6 suggest that by subtracting polynomials of higher and higher degree from $u(x, t)$, the residual Fourier series can be made to converge faster and faster. Subtracting a linear polynomial as in (6.15) gives Fourier approximations with errors of order $N^{-3/2}$ as $N\to\infty$; subtracting a quadratic polynomial gives Fourier approximations with errors of order $N^{-7/2}$; and so on. In the limit we dispense entirely with Fourier series and obtain a rapidly converging polynomial approximation. The convergence theory of these polynomial spectral approximations is discussed in the next two sections.

SECTION 7

Applications of Algebraic-Stability Analysis

The main result of § 5 does not provide us with a systematic way of constructing the family H_N of Lyapunov matrices necessary to prove algebraic stability. In general, these matrices are difficult to find. However, there are several problems for which they can be found directly from the differential equation.

It is very easy to construct Lyapunov matrices for Galerkin approximations to

$$\frac{\partial u}{\partial t}=Lu$$

when L is a semi-bounded operator on the Hilbert space $\mathcal{H}$. We say that L is semi-bounded if

$$L+L^* \leqq \alpha I \tag{7.1}$$

for some constant α, where L^* is the adjoint of L defined with respect to the Hilbert space inner product $(\cdot\,,\cdot\,)$. If L is semi-bounded

$$\frac{d}{dt}(u, u) \leqq \alpha(u, u), \tag{7.2}$$

so

$$(u(t), u(t)) \leqq e^{\alpha t}(u(0), u(0))$$

and the "energy" $(u(t), u(t))$ grows at most exponentially with t.

If an energy estimate of the form (7.2) exists, then Galerkin approximation based on the Hilbert space inner product $(\cdot\,,\cdot\,)$ is stable (and, hence, algebraically stable). The Lyapunov matrix H_N may be chosen to be the $N \times N$ identity matrix I_N. It follows from the Galerkin equations (2.7)–(2.8) with $f \equiv 0$ that

$$\frac{d}{dt}(u_N, u_N)=(u_N, [L+L^*]u_N) \leqq \alpha(u_N, u_N).$$

Thus,

$$(u_N(t), u_N(t)) \leqq e^{\alpha t}(u_N(0), u_N(0)).$$

Since $u_N(t)=\exp(L_N t)u_N(0)$ for all $u_N(0)$, we obtain $\|\exp(L_N t)\| \leqq \exp(\frac{1}{2}\alpha t)$ so stability is proved. The reader is reminded that with stability established, the theory of §§ 4 and 5 proves convergence for *consistent* schemes.

EXAMPLE 7.1. *Semi-bounded Galerkin approximations.* The above construction establishes stability and thus convergence for a wide variety of Galerkin approximations. Among these stable Galerkin approximations are:

(i) Solution of any problem $u_t = Lu$ that is semi-bounded in L_2 $(-1, 1)$ by means of Legendre series. For example, $u_t + u_x = f(x, t)$ with $u(-1, t) = 0$ is stable (and convergent) when solved by Legendre–Galerkin approximation. For our argument to be complete it is necessary to verify that the Legendre–Galerkin approximation to this problem is consistent. This is done as follows.

We write

$$\|Lu - P_N L P_N u\| \leqq \|(I - P_N)Lu\| + \|P_N L(I - P_N)u\|.$$

The first term on the right goes to zero as $N \to \infty$ at a rate governed solely by the smoothness of Lu; it measures the error in the N term Legendre–Galerkin expansion of Lu. The second term is estimated as follows. Set

$$L(I - P_N)u = \sum_{n=1}^{\infty} a_n \phi_n(x)$$

where $\{\phi_n\}$ are normalized Legendre polynomials. If L is a finite-order differential operator so L^* is also a finite-order differential operator (for example, $L^* = \partial/\partial x$ if $L = -\partial/\partial x$), then

$$\begin{aligned} a_n &= (\phi_n, L(I - P_N)u) \\ &= (L^* \phi_n, (I - P_N)u). \end{aligned}$$

Thus,

$$\begin{aligned} |a_n| &\leqq \|L^* \phi_n\| \|(I - P_N)u\| \\ &= O(n^A / N^B), \qquad n \to \infty, \quad N \to \infty, \end{aligned}$$

where A depends only on L ($A = 3/2$ if $L = -\partial/\partial x$ and ϕ_n is a normalized Legendre polynomial) and B depends only on the smoothness of u (B is arbitrary if u is infinitely differentiable). Thus,

$$\|P_N L(I - P_N)u\|^2 = \sum_{n=1}^{N} a_n^2 \to 0$$

faster than any power of $1/N$ as $N \to \infty$ if u and all its derivatives are smooth. This proves consistency. This kind of proof extends to a wide variety of the examples to be discussed in §§ 7 and 8, but will not be repeated.

(ii) Solution of $u_t = x u_x$ with the boundary conditions $u(\pm 1, t) = 0$ is a well posed problem in the Chebyshev inner product

$$(u, v) = \int_{-1}^{1} \frac{u(x)v(x)}{(1 - x^2)^{1/2}} \, dx.$$

In fact, if $L = x\, \partial/\partial x$, u is differentiable, and $u(\pm 1) = 0$, then, by integration by parts,

$$(u, Lu) = \int_{-1}^{1} x(1 - x^2)^{-1/2} u u_x \, dx = -\int_{-1}^{1} (1 - x^2)^{-3/2} u^2 \, dx \leqq 0.$$

Thus, Galerkin approximation to the problem is stable using Chebyshev polynomials.

(iii) Solution of $u_t+u_x=0$ $(0\leqq x<\infty)$ with $u(0,t)=0$ is a well posed problem in the Laguerre inner product

$$(u,v)=\int_0^\infty u(x)v(x)\,e^{-x}\,dx.$$

In fact, if $u(0,t)=0$ then, by integrating by parts,

$$-\int_0^\infty uu_x\,e^{-x}\,dx=-\tfrac{1}{2}e^{-x}u^2\Big|_0^\infty-\int_0^\infty e^{-x}u^2\,dx<0.$$

Similarly, the problem $u_t=u_{xx}$ $(0\leqq x<\infty)$ with $u(0,t)=0$ is also stable in the Laguerre norm.

(iv) Solution of $u_t=-xu_x$ $(-\infty<x<\infty)$ is well posed in the Hermite inner product

$$(u,v)=\int_{-\infty}^\infty u(x)v(x)\,e^{-x^2}\,dx.$$

In fact,

$$\frac{\partial}{\partial t}(u,u)=-2\int_{-\infty}^\infty x\,e^{-x^2}uu_x\,dx,$$

so that integration by parts gives

$$\frac{\partial}{\partial t}(u,u)=\int_{-\infty}^\infty u^2\,e^{-x^2}(1-2x^2)\,dx\leqq(u,u),$$

where we assume that $u\ll x^{-1/2}\exp(\frac{1}{2}x^2)$ as $|x|\to\infty$.

(v) The heat equation $u_t=u_{xx}$ with $u(\pm1,t)=0$ is semibounded in the Chebyshev norm. In fact, if u is differentiable for $|x|\leqq 1$ then

$$\int_{-1}^1(1-x^2)^{-1/2}uu_{xx}\,dx=(1-x^2)^{-1/2}uu_x\big|_{-1}^1-\int_{-1}^1[u(1-x^2)^{-1/2}]_xu_x\,dx.$$

The first term vanishes because u is a polynomial in x and therefore $u(\pm1)=0$ implies

$$\frac{u}{(1-x^2)^{1/2}}\bigg|_{x=\pm1}=0.$$

The integral term on the right is

$$
\begin{aligned}
-\int_{-1}^{1}[u(1-x^2)^{-1/2}]_x u_x\,dx &= -\int_{-1}^{1}[u(1-x^2)^{-1/2}]_x[u(1-x^2)^{-1/2}]_x(1-x^2)^{1/2}\,dx \\
&\quad+\tfrac{1}{2}\int_{-1}^{1}\frac{\partial}{\partial x}[u(1-x^2)^{-1/2}]^2 x(1-x^2)^{-1/2}\,dx \\
&= -\int_{-1}^{1}[(u(1-x^2)^{-1/2})_x]^2(1-x^2)^{1/2}\,dx \\
&\quad+\tfrac{1}{2}u^2x(1-x^2)^{-3/2}\Big|_{-1}^{1}-\tfrac{1}{2}\int_{-1}^{1}u^2(1-x^2)^{-5/2}\,dx \\
&\leqq 0,
\end{aligned}
\tag{7.3}
$$

and therefore,

$$\frac{d}{dt}\int_{-1}^{1}\frac{u^2}{\sqrt{1-x^2}}\,dx \leqq 0.$$

In the next three examples we generalize the proofs of stability and convergence for Galerkin approximations given in Example 7.1 to show the stability and convergence of tau approximations.

EXAMPLE 7.2. *A semi-bounded tau approximation.* Consider the equation

$$\frac{\partial u}{\partial t}=x\frac{\partial u}{\partial x}$$

with the boundary conditions

$$u(\pm 1, t)=0.$$

It was shown in Example 7.1(ii) that if $L=x\partial/\partial x$, then

$$L+L^*\leqq 0$$

in the Chebyshev inner product. If we seek the solution as the truncated Chebyshev series

$$u_N=\sum_{n=0}^{N}a_nT_n$$

by the tau method, then u_N satisfies exactly the equation

$$\frac{\partial u_N}{\partial t}-x\frac{\partial u_N}{\partial x}=\tau_N(x)T_N(x)+\tau_{N-1}(t)T_{N-1}(x). \tag{7.4}$$

Equating coefficients of x^N and x^{N-1} on both sides of (7.4), we obtain

$$a'_N - Na_N = \tau_N,$$

$$a'_{N-1} - (N-1)a_{N-1} = \tau_{N-1}$$

since $T_n = 2^{n-1}x^n - n2^{n-3}x^{n-2} + \cdots$. Therefore,

$$\frac{2}{\pi}\left(u_N, \frac{\partial u_N}{\partial t}\right) = \frac{1}{\pi}([L+L^*]u_N, u_N) + [a'_N - Na_N]a_N + [a'_{N-1} - (N-1)a_{N-1}]a_{N-1} \tag{7.5}$$

so that

$$\frac{1}{2}\frac{\partial}{\partial t}\left[\frac{2}{\pi}(u_N, u_N) - a_N^2 - a_{N-1}^2\right] = \frac{1}{\pi}([L+L^*]u_N, u_N) - Na_N^2 - (N-1)a_{N-1}^2 \leqq 0.$$

Since

$$(u_N, u_N) = \frac{\pi}{2}\sum_{n=0}^{N} c_n a_n^2$$

with $c_0 = 2$, $c_n = 1$ $(n \geqq 1)$, the above inequality is equivalent to

$$\frac{\partial}{\partial t}\sum_{n=0}^{N-2} c_n a_n^2 \leqq 0. \tag{7.6}$$

This proves stability: a_N and a_{N-1} are bounded because they are determined in terms of $a_0, a_1, \cdots, a_{N-2}$ by the boundary conditions $u(\pm 1, t) = 0$.

For this example, we can prove stability directly from the matrix representation of L_N. In fact,

$$(L_N)_{jk} = \frac{1}{c_j}\left[-k\delta_{jk} + 2\sum_{\substack{l=0 \\ l \text{ even}}}^{N-j} k\delta_{j+l,k}\right], \qquad 0 \leqq j \leqq N, \quad 0 \leqq k \leqq N, \tag{7.7a}$$

In the tau approximation, the boundary conditions $u(\pm 1, t) = 0$ require that the last two rows of the matrix L_N be replaced by

$$(L_N)_{N-1,k} = (-1)^k, \tag{7.7b}$$

$$(L_N)_{N,k} = 1. \tag{7.7c}$$

If the boundary conditions (7.7b,c) are not applied then the spectral approximation is unstable: without the boundary conditions L_N has the eigenvalue N [with the eigenvector $a_{N-2k} = \binom{N}{k}$, $a_{N-2k-1} = 0$] so that

$$\|e^{L_N t}\| \geqq e^{Nt}.$$

To prove convergence when the boundary conditions (7.7b,c) are applied, let us only consider an odd solution in which $a_n = 0$ if n is even. If we assume that $N = 2M+1$ and set

$$d_k = a_{2k+1}, \qquad 0 \leqq k \leqq M,$$

then the system reduces to

$$\frac{\partial \mathbf{d}}{\partial t} = D\mathbf{d},$$

where

$$D_{jk} = -(2k+1)\delta_{jk} + 2\sum_{l=0}^{M-j} (2k+1)\delta_{j+l,k} - 2N,$$

$$0 \leqq j < M, \quad 0 \leqq k < M.$$

If we introduce the $M \times M$ transformation matrix S defined by

$$S_{jk} = \delta_{jk} - \delta_{j+1,k}, \qquad 0 \leqq j < M, \quad 0 \leqq k < M,$$

then $S(D+D^*)S^*$ is a diagonal matrix with entries $(-4, -4, \cdots, -4N-12)$. Thus we obtain $D + D^* \leqq 0$, so that $\partial(\mathbf{d}, \mathbf{d})/\partial t \leqq 0$, which proves stability.

EXAMPLE 7.3. *Stability of tau methods applied to degree reducing semi-bounded equations.* An argument similar to that given in Example 7.2 demonstrates stability of tau methods in terms of arbitrary orthonormal polynomial bases for equations $\partial u/\partial t = Lu$ where L is semi-bounded and degree reducing: L is said to be *degree reducing* if for any polynomial P_N of degree N, LP_N is a polynomial of degree at most $N-k$ where k is the number of boundary conditions that are applied. If L is degree reducing, equating coefficients of $x^{N-k+1}, \cdots, x^N$ in

$$\frac{\partial u_N}{\partial t} = Lu_N + \sum_{n=N-k+1}^{N} \tau_n \phi_n$$

implies that $\tau_n(t) = a_n'(t)$ for $n = N-k+1, \cdots, N$; here

$$u_N(x, t) = \sum_{n=0}^{N} a_n(t)\phi_n(x)$$

and the orthonormal expansion polynomial $\phi_n(x)$ is assumed to be of degree n. Therefore,

$$\frac{1}{2}\frac{\partial}{\partial t}(u_N, u_N) - \sum_{n=N-k}^{N} a_n' a_n = ([L+L^*]u_N, u_N) \leqq 0$$

so that

$$\frac{\partial}{\partial t}\sum_{n=0}^{N-k} a_n^2 \leqq 0,$$

which proves stability since $a_{N-k+1}, \cdots, a_N$ are determined by the boundary conditions in terms of $a_0, a_1, \cdots, a_{N-k}$.

EXAMPLE 7.4. *More stable tau approximations.* (i) Suppose that

$$u_t + u_x = 0, \qquad -1 \leqq x \leqq 1, \quad t > 0,$$

$$u(-1, t) = 0$$

is solved by tau approximation using Legendre polynomials. The Nth degree Legendre–tau approximation u_N satisfies

$$\frac{\partial}{\partial t}u_N+\frac{\partial}{\partial x}u_N=a'_NP_N, \qquad u_N(-1,t)=0.$$

Since $\int_{-1}^{1} P_N^2(x)\,dx=2/(2N+1)$ [see (A.25)], we obtain

$$\frac{d}{dt}\Big[\int_{-1}^{1} u_N^2\,dx-\frac{2}{2N+1}a_N^2\Big]\leqq 0,$$

which proves stability.

(ii) Suppose that

$$u_t=u_{xx},$$
$$u(\pm 1,t)=0$$

is solved by the tau method using Chebyshev polynomials. Since $L=\partial^2/\partial x^2$ is degree reducing and $L+L^*\leqq 0$ [see Example 7.1(v)), the method is stable.

(iii) The solution of

$$\begin{aligned} &u_t+u_x=0 && 0\leqq x\leqq\infty,\quad t>0,\\ &u(0,t)=\sin t, && t>0,\\ &u(x,0)=0, && 0\leqq x\leqq\infty, \end{aligned} \tag{7.8}$$

by Laguerre polynomials is stable using the tau method since, by Example 7.1(iii), L is semi-bounded. The equations of the Laguerre–tau approximation to (7.8) are a simple modification of (2.24)–(2.25). In Fig. 7.1 we compare this tau approximation with the exact solution of (7.8) at $t=30$ for a 20-term Laguerre expansion. The reader should compare this approximate result obtained by the tau method with the best Laguerre approximation to sin x plotted in Fig. 3.12.

In the next example we discuss some ways to find nontrivial Lyapunov matrices $\{H_N\}$ when L is not semi-bounded.

EXAMPLE 7.5. *Polynomial approximations to a variable coefficient hyperbolic equation.* Consider the initial-value problem

$$\begin{aligned} &u_t=-xu_x, && |x|<1,\\ &u(x,0)=g(x), \end{aligned} \tag{7.9}$$

which is well posed without requiring any boundary conditions. The exact solution to this problem is

$$u(x,t)=g(x\,e^{-t})$$

so that $u(x,t)$ approaches a constant as $t\to\infty$:

$$u(x,t)\sim g(0), \qquad t\to\infty,\quad |x|\leqq 1.$$

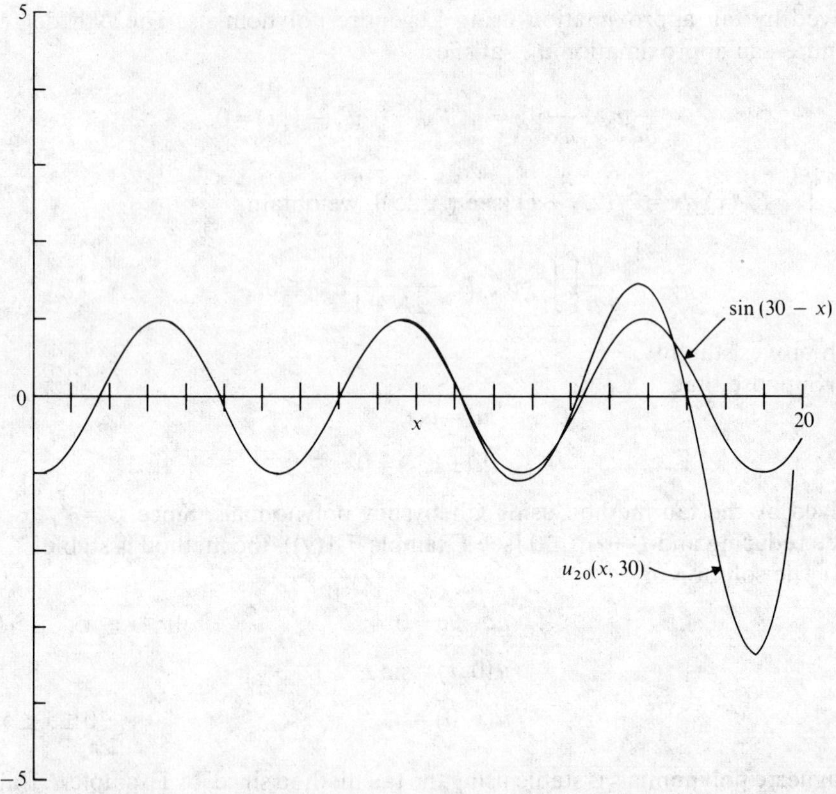

FIG. 7.1. *A plot of the Laguerre–tau approximation $u_N(x, t)$ to the problem (7.8) and a plot of the exact solution to this problem for $t = 30$, $N = 20$. Here the Laguerre expansion (2.23) is truncated at $N = 20$ and the tau equations are given by (2.24) with $f = 0$ and (2.25) replaced by $\sum_{n=0}^{N} a_n = \sin t$. Observe the similarity of the Laguerre–tau approximation to the best Laguerre approximation to $u(x, 30)$ plotted in Fig. 3.12.*

The problem is well posed in the sense that $\|\exp(Lt)\|$ is finite for finite t, where $L = -x\partial/\partial x$ and $\|\cdot\|$ is the usual L_2 norm. However, $\|\exp(Lt)\| = \exp(\frac{1}{2}t)$ so $\|\exp(Lt)\|$ is unbounded as $t \to \infty$. To show this, we observe that the function that extremizes $\|u(t)\|$ subject to $\|u(0)\| = 1$ satisfies $u(x, 0) = g_t(x)$ where

$$g_t(x) = \begin{cases} \pm \dfrac{e^{t/2}}{\sqrt{2}}, & |x| \leqq e^{-t}, \\ 0, & |x| > e^{-t}. \end{cases}$$

The operator L is semibounded in the usual L_2 norm:

$$\frac{\partial}{\partial t} \int_{-1}^{1} u^2\, dx = -\int_{-1}^{1} x \frac{\partial u^2}{\partial x}\, dx = -u^2(1) - u^2(-1) + \int_{-1}^{1} u^2\, dx$$

$$\leqq \int_{-1}^{1} u^2\, dx,$$

so $L+L^* \leqq I$. Therefore, Galerkin polynomial solution of (7.9) is stable and convergent. The Legendre polynomial approximation $u_N(x, t)$ satisfies

$$\frac{\partial u_N}{\partial t}+x\frac{\partial u_N}{\partial x}=0 \tag{7.10}$$

exactly because no boundary conditions are applied and L is degree preserving. Therefore, Galerkin, tau, and collocation approximations to (7.9) are *identical* and all three methods are stable.

In fact, all polynomial-spectral methods applied to (7.9) satisfy (7.10); all polynomial methods for this problem give identical results and, therefore, they are all stable in the usual L_2 norm. In terms of the natural norms for a general polynomial basis $\{\psi_n\}$, i.e., that norm in which $(\psi_i, \psi_j)=\delta_{ij}$, the spectral approximation (7.10) is algebraically stable if the $N\times N$ matrix whose elements are

$$(H_N)_{jk} = \int \psi_j(x)\psi_k(x)\,dx$$

has a condition number which is bounded algebraically, i.e., $\|H_N\|\|H_N^{-1}\|=O(N^\beta)$ $(N\to\infty)$.

As an example of the complicated behavior of spectral approximations for this problem in norms different from the usual L_2 norm, let us consider the Chebyshev–L_2 norm. It may easily be shown that $L+L^*$ is not semibounded in the Chebyshev inner product. For example, consider the trial function

$$v=T_0-T_{2N};$$

we obtain

$$\begin{aligned}([L+L^*]v, v) &= -(xv_x, v)-(v, xv_x)\\ &= -\left(-2N[T_{2N-1}+\cdots+T_1],\, T_1-\frac{T_{2N+1}+T_{2N-1}}{2}\right)\\ &= \tfrac{1}{3}N(v, v).\end{aligned}$$

Nevertheless, Chebyshev approximation to this problem is algebraically stable. We will demonstrate this fact explicitly by construction of a Lyapunov matrix.

A Lyapunov matrix for the Chebyshev approximation to (7.9) may be found by direct examination of the evolution equation for the vector $\mathbf{a}_N=(a_0,\cdots, a_N)$:

$$\frac{\partial a_n}{\partial t}=-na_n-2\sum_{\substack{p=n+2\\ p+n \text{ even}}}^{N} pa_p, \qquad n=0,\cdots, N. \tag{7.11}$$

Since a_0 decouples from $a_1,\cdots, a_N$ in (7.11), we can restrict attention to $a_1,\cdots, a_N$. Suppose we define the matrix H_N by

$$(H_N)_{jk}=\frac{1}{j}\,\delta_{jk}, \qquad 1\leqq j, k\leqq N.$$

H_N is the Lyapunov matrix for L_N^T; in fact (5.7b) is satisfied because

$$\tfrac{1}{2}(H_N L_N + L_N^T H_N) = \begin{pmatrix} -1 & 0 & -1 & 0 & -1 & \cdots \\ 0 & -1 & 0 & -1 & 0 & \cdots \\ -1 & 0 & -1 & 0 & -1 & \cdots \\ 0 & -1 & 0 & -1 & 0 & \cdots \\ \vdots & \vdots & \vdots & \vdots & \vdots & \ddots \end{pmatrix} \leqq 0.$$

The matrix displayed above has rank 2 and the nonzero eigenvalues are $-[N/2]$, $-[(N+1)/2]$. Therefore, by the theory of § 5,

$$\|e^{L_N t}\| \leqq \sqrt{\|H_N\| \|H_N^{-1}\|} = \sqrt{N},$$

where $\|\cdot\|$ is now the Chebyshev norm. Thus, L_N is algebraically stable in the Chebyshev norm even though $L_N + L_N^*$ is unbounded in this norm.

The qualitative behavior of the Chebyshev norm of $\exp(L_N t)$ as a function of N and t is as follows. For fixed t and $N \to \infty$, $\|\exp(L_N t)\| = O(N^{1/4})$; this result is justified heuristically by following the argument given in § 5 that led to (5.4). On the other hand if $t \gtrsim \ln N$, $\|\exp(L_N t)\| = O(N^{1/2})$ as $N \to \infty$. A heuristic justification of this result is as follows. Let $u(x, 0) = 1$ for $|x| \leqq \varepsilon$, 0 for $|x| > \varepsilon$. Then the exact solution of (7.9) for $t > \ln 1/\varepsilon$ is $u(x, t) \sim 1$ for $|x| \leqq 1$, so $\|u(x, t)\|^2 \sim \pi$ as $\varepsilon \to 0^+$ for $t > \ln 1/\varepsilon$. As in § 5, we conclude that $\|\exp(L_N t)\| = O(N^{1/2})$ for $t \gtrsim \ln N$ as $N \to \infty$. [Even in the usual L_2 norm, $\|\exp(L_N t)\| = O(N^{1/2})$ when $t \gtrsim \ln N$, which mimics the unbounded growth of $\|\exp(Lt)\|$ as $t \to \infty$.]

SECTION 8

Constant Coefficient Hyperbolic Equations

In this section, we discuss the stability of spectral methods for the problem

(8.1) $$\frac{\partial u}{\partial t}+\frac{\partial u}{\partial x}=0, \qquad |x|\leqq 1, \quad t>0,$$

with the initial condition

(8.2) $$u(x,0)=f(x), \qquad |x|\leqq 1,$$

and the boundary condition

(8.3) $$u(-1,t)=0, \qquad t>0.$$

The results for this problem can be extended to a general hyperbolic system of the form

$$u_t = Au_x$$

with characteristic boundary conditions, because for any hyperbolic system A can be diagonalized by a real similarity transformation.

The operator $L=-\partial/\partial x$ is semi-bounded in the usual $L_2(-1,1)$ norm when operating on the subspace of functions v that satisfy the boundary condition $v(-1,t)=0$. In fact

$$(v,[L+L^*]v)=-2\int_{-1}^{1} v\frac{\partial v}{\partial x}\,dx=-v^2(1)\leqq 0$$

and therefore Galerkin and tau methods are stable using Legendre polynomials.

However, L is not semi-bounded in the Chebyshev norm. To show this, we set

$$v(x)=T_{2N}(x)-T_1(x)-2T_0(x)$$

so that $v(-1)=0$. Since

$$T'_{2N}=2N[T_{2N-1}+T_{2N-3}+\cdots+T_1],$$

we find

(8.4)
$$\begin{aligned}(v,[L+L^*]v) &= -2\int_{-1}^{1}(1-x^2)^{-1/2}\frac{\partial v}{\partial x}v\,dx\\ &= -2\int_{-1}^{1}(1-x^2)^{-1/2}[2N(T_{2N-1}+T_{2N-3}+\cdots+T_1)-T_0]\\ &\qquad\cdot(T_{2N}-T_1-2T_0)\,dx\\ &\doteq \frac{2N-4}{5}(v,v).\end{aligned}$$

The fact that $L+L^*$ is not semi-bounded is consistent with the fact that $\exp(Lt)$ is not a bounded operator for $t<2$ in the Chebyshev norm (see § 5). However, these results do not prove that Chebyshev–spectral approximation to (8.1)–(8.3) is not convergent. In fact, we shall show that, while Chebyshev–spectral approximation to (8.1)–(8.3) is not stable in the Chebyshev–L_2 norm, it is algebraically stable in this norm.

In order to investigate algebraic stability, we must study more carefully the behavior of the Chebyshev coefficients of the approximate solution

$$u_N=\sum_{n=0}^{N} a_n(t)T_n(x).$$

The differential equations for the a_n's are given by (2.12) for Galerkin approximation, (2.20) for the tau method, and (2.33) for the collocation method. As remarked in § 2, all these equations may be written in the vector form

$$\frac{\partial \mathbf{a}}{\partial t}=L_N\mathbf{a}$$

where $\mathbf{a}=(a_0, a_1, \cdots, a_{N-1})$ and L_N is an $N\times N$ matrix. The value of $a_N(t)$ is determined in terms of $\mathbf{a}(t)$ by the boundary condition (8.3).

Numerical evidence for algebraic stability. Let us first examine the behavior of $L_N+L_N^*$. In Table 8.1 we list the largest eigenvalue of $L_N+L_N^*$ for $N=10, 20, \cdots, 100$ for the three Chebyshev methods. This table indicates that the largest positive eigenvalue of $L_N+L_N^*$ grows like cN^2 for some constant c. If L_N were a normal matrix this would imply that $\|e^{L_Nt}\|$ behaves like $\exp(\frac{1}{2}cN^2t)$.

TABLE 8.1

The largest positive eigenvalue $\lambda_{\max}$ of $L_N+L_N^*$ for the Chebyshev–spectral solution of the one-dimensional wave equation (8.1)–(8.3). *The Galerkin approximation to this problem is given by the solution to* (2.12), *the tau approximation is given by* (2.20), *and the collocation approximation is given by* (2.33). *Observe that $\lambda_{\max}\sim cN^2$ as $N\to\infty$ where $c \doteqdot 0.13$ for the Galerkin and collocation methods and $c \doteqdot 0.21$ for the tau method.*

N	Galerkin	Tau	Collocation
10	12.8945	21.4090	13.6382
20	51.7629	84.8970	52.5702
30	116.1687	190.4908	117.0084
40	206.1141	338.1769	206.9735
50	321.5992	527.9525	322.4719
60	462.6240	759.8167	463.5062
70	629.1885	1033.7691	630.0778
80	821.2926	1349.8094	822.1875
90	1038.9364	1707.9375	1039.8357
100	1282.1199	2108.1535	1283.0228

However, the matrices L_N are not normal and therefore the large eigenvalues of $L_N+L_N^*$ do not imply instability.

TABLE 8.2

The largest eigenvalue λ_{max} of $\exp(L_N)\exp(L_N^*)$. The L_2 matrix norm of $\exp(L_N)$ is related to λ_{max} by $\|\exp(L_N)\|^2=\lambda_{max}$. As $N\to\infty$, λ_{max} behaves like cN with $c \doteq 0.8$ for the Galerkin and collocation methods, while λ_{max} behaves like $cN^{1/2}$ with $c \doteq 0.6$ for the tau method. Observe that the largest eigenvalues of $\exp(L_N)\exp(L_N^*)$ grow at most like N despite the existence of eigenvalues of $L_N+L_N^*$ that grow like N^2 (see Table 8.1).

N	Galerkin	Tau	Collocation
10	8.0362	2.0004	7.9744
20	15.5018	2.8120	15.3995
30	22.9485	3.4857	22.9908
40	30.6542	4.0514	30.5421
50	37.9723	4.5339	37.9493
60	45.5068	4.9856	45.5133
70	53.1338	5.4002	53.0671
80	60.5130	5.7771	60.4705
90	67.9925	6.1402	68.0282
100	75.6644	6.4831	75.5778

In Table 8.2, we give the norms of the matrices $\exp[L_N]\cdot\exp[L_N^*]$ for the three projection methods (Galerkin, collocation, and tau). The results indicate that $\|\exp(L_N)\|$ grows only like $N^{1/2}$ as $N\to\infty$ for all three methods. In other words, L_N is algebraically stable (at least for $t=1$). This result shows the extreme pessimism of the energy estimate $\|\exp(L_N)\|=O(\exp(\frac{1}{2}cN^2))$; crude energy methods may be very misleading for nonnormal evolution operators.

In order to understand better how the Chebyshev–spectral methods avoid an energy "catastrophe" (energy growth like $\exp(cN^2t)$) we have solved the tau equations (2.20) numerically with a very "bad" initial condition:

$$u_N(x,0)=[T_N(x)+2T_{N-1}(x)+(-1)^N T_0(x)]/\sqrt{7}. \tag{8.5}$$

For the tau method, this initial condition satisfies

$$\frac{\partial}{\partial t}(u_N,u_N)\bigg|_{t=0}=(u_N,(L_N+L_N^*)u_N)=O(N^2), \qquad N\to\infty.$$

In Figs. 8.1–8.2 we plot the energy (u_N,u_N) vs. t for $N=25$ and $N=50$. It is apparent that the initial slope of the energy growth is of order N^2 but that the energy does not maintain this rapid rate of growth. Observe that the region of rapid growth is closer to $t=0$ for $N=50$ than for $N=25$. The behavior observed in Figs. 8.1–8.2 is not inconsistent with the fact that $u_N(t=0)$ is a "bad" eigenmode of $L_N+L_N^*$. Because L_N is nonnormal the "bad" initial condition is not an eigenmode of L_N so that after evolution from 0 to t by $\exp(L_Nt)$, u_N "rotates" out of the region of bad modes of $L_N+L_N^*$.

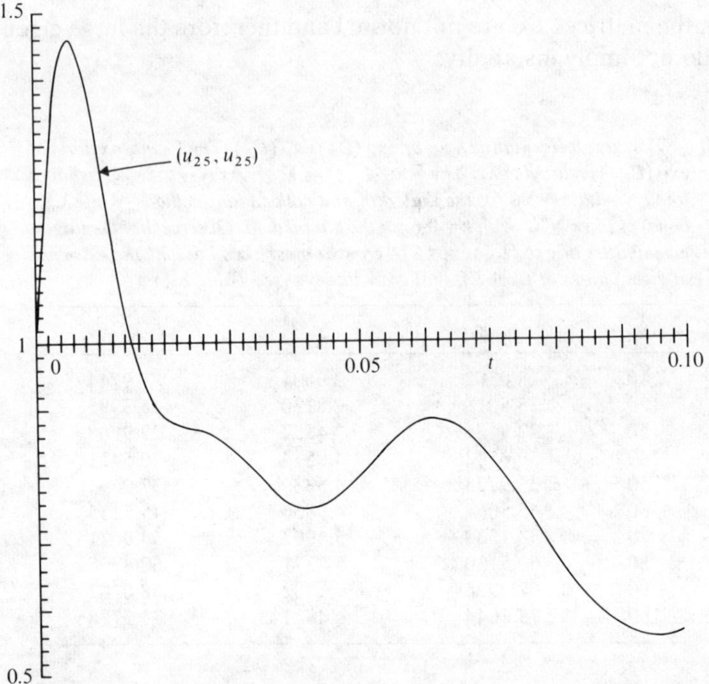

FIG. 8.1. *A plot of the energy* (u_N, u_N) *vs.* t *for the "bad" initial conditions* (8.5) *with* $N=25$. *Here* $u_N(x, t)$ *is the solution to the Chebyshev–tau equation* (2.20) *for* $u_t+u_x=0$ *with* $N=25$, $u_{25}(x, 0)=(T_{25}+2T_{24}-T_0)/\sqrt{7}$.

The direct computation of $\exp[L_N t]$ for $t=1$ is not enough to verify algebraic stability because the theory of § 5 shows that we must study the behavior of $\exp[L_N t]$ for a complete time interval $0 \leqq t \leqq T$. This may be done using the method suggested in § 5 for the numerical verification of algebraic stability. First, in Table 8.3 we list the numerically computed eigenvalues of L_N. Observe that all the eigenvalues of L_N have negative real part. (This result will be shown rigorously later.) Therefore, $\|\exp(L_N t)\| \to 0$ as $t \to \infty$ for fixed N. Thus the Chebyshev approximations are asymptotically stable in the sense that they remain bounded as $t \to \infty$ with N fixed.

In Figs. 8.3–8.5, we plot the L_1 matrix norm of $\exp(L_N t)$ vs. t in the tau method for $N=5, 15, 25$. Observe that as $t \to \infty$ for fixed N, $\|\exp(L_N t)\|_1$ approaches zero while it grows slowly (like $N^{1/2}$) as $N \to \infty$ for fixed $t<2$. (Note that growth of $\|\exp(L_N t)\|_1$ like $N^{1/2}$ as $N \to \infty$ is not inconsistent with growth of $\|\exp(L_N t)\|_2$ like $N^{1/4}$.) Also observe that the norms seem to have a boundary layer at $t=2$ such that $\|\exp(L_N t)\|_1 \to \infty$ as $N \to \infty$ for $t<2$ and $\to 0$ as $N \to \infty$ for $t>2$. This behavior is consistent with the unboundedness of $\exp(Lt)$ for $t<2$ [see (5.4)].

Asymptotic stability does not prove *stability* because L_N is not normal. The next step in the computational proof of stability is to compute numerically the Lyapunov matrices H_N satisfying

$$H_N A_N + A_N^* H_N = -I. \tag{8.6}$$

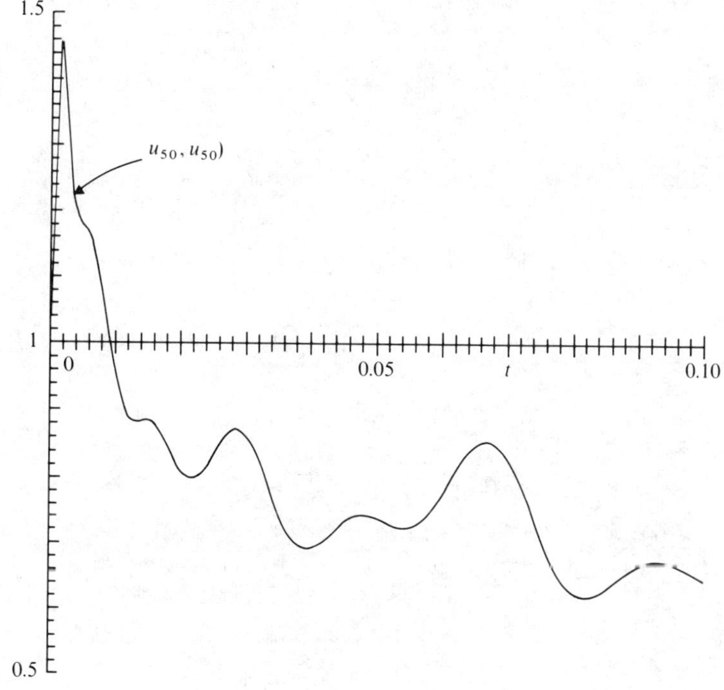

FIG. 8.2. *Same as Fig.* 8.1 *except for* $N=50$ *with* $u_{50}(x, 0)=(T_{50}+2T_{49}+T_0)/\sqrt{7}$.

TABLE 8.3

The real part of the eigenvalue of L_N *with least negative real part for the collocation, tau, and Galerkin spectral approximations to* (8.1)–(8.3). *Since all the eigenvalues of* L_N *have negative real parts, these spectral methods are asymptotically stable as* $t \to \infty$.

N	Galerkin	Tau	Collocation
10	−2.4532	−2.9994	−1.9306
20	−2.5932	−3.9320	−2.1591
30	−2.7267	−4.5380	−2.3247
40	−2.8495	4.9910	−2.4659
50	−2.9669	−5.3837	−2.5965
60	−3.0824	−5.7266	−2.7226
70	−3.1985	−6.0489	−2.8478
80	−3.3162	−6.3650	−2.9738
90	−3.4365	−6.6861	−3.1017
100	−3.5597	−7.0229	−3.4335

A good method to compute H_N is described by Bartels and Stewart (1972). In Table 8.4 we list the condition number of H_N for the Galerkin, collocation and tau methods. This table suggests that the condition number of H_N grows at most like

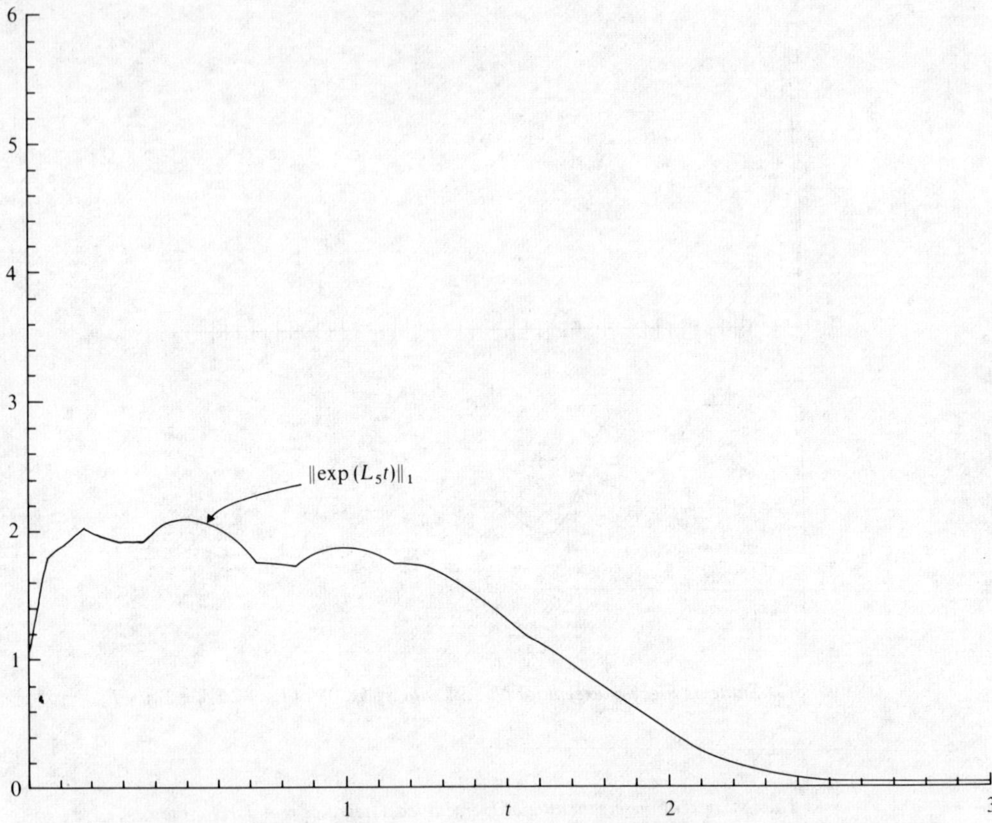

FIG. 8.3. *A plot of the L_1 matrix norm of* exp $(L_N t)$ *vs. t where L_N is the Chebyshev–tau approximation to* (8.1)–(8.3). *Here $N=5$ so the Chebyshev polynomial expansion is truncated after $T_5(x)$.*

N^3 as N for the Galerkin and collocation methods[3] and like N^2 for the tau method. Recalling (5.11), we obtain

$$\|[\exp(L_N t)]\| = O(N^{3/2} e^{-t/2}). \tag{8.7}$$

for all three methods. It should be noted that (8.7) gives only an upper bound for $\|\exp[L_N t]\|$. According to the theory given in § 5, this upper bound can be sharpened by at most $\|L_N\| = O(N^2)$ $(N \to \infty)$, explaining the origin of the difference between the estimate (8.7) and the observed behavior $N^{1/2}$ of the computed L_2 matrix norms.

In the above discussion, we have given numerical evidence for algebraic stability of the Chebyshev–spectral methods for (8.1). We shall now prove rigorously that Chebyshev–spectral methods for (8.1) are algebraically stable.

---

[3] The condition number of H_N can grow no faster than N^3 as $N \to \infty$. To see this, we note that (5.14) gives $\|H_N^{-1}\| = O(N^2)$ while (5.13) and the results that $\|\exp(L_N t)\| = O(N^{1/2})$ for $t \leqq 2$ and $\|\exp(L_N t)\| \to 0$ as $N \to \infty$ for $t > 2$ give $\|H_N\| = O(N)$ as $N \to \infty$.

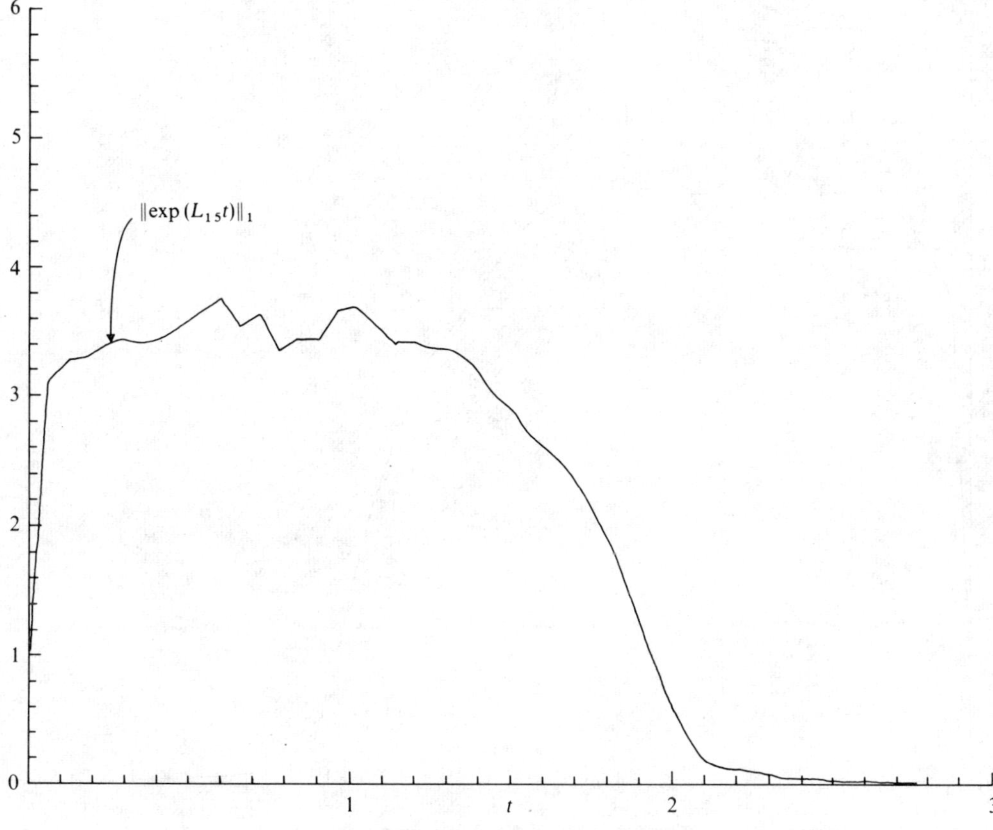

FIG. 8.4. *Same as Fig. 8.3 except* $N = 15$.

Proof of algebraic stability for Chebyshev–Galerkin approximation. In the Chebyshev–Galerkin approximation to (8.1), we represent the spectral approximation u_N by the series

$$u_N = \sum_{n=1}^{N} a_n(t)[T_n - (-1)^n T_0]. \tag{8.8}$$

Recalling (2.35), u_N satisfies

$$\frac{\partial u_N}{\partial t} + \frac{\partial u_N}{\partial x} = \tau_N(t) \sum_{n=0}^{N} \frac{T_n(x)}{c_n}(-1)^n. \tag{8.9}$$

We can determine $\tau_N(t)$ by equating the coefficients of x^N in (8.9):

$$\tau_N(t) = \frac{da_N(t)}{dt}(-1)^N.$$

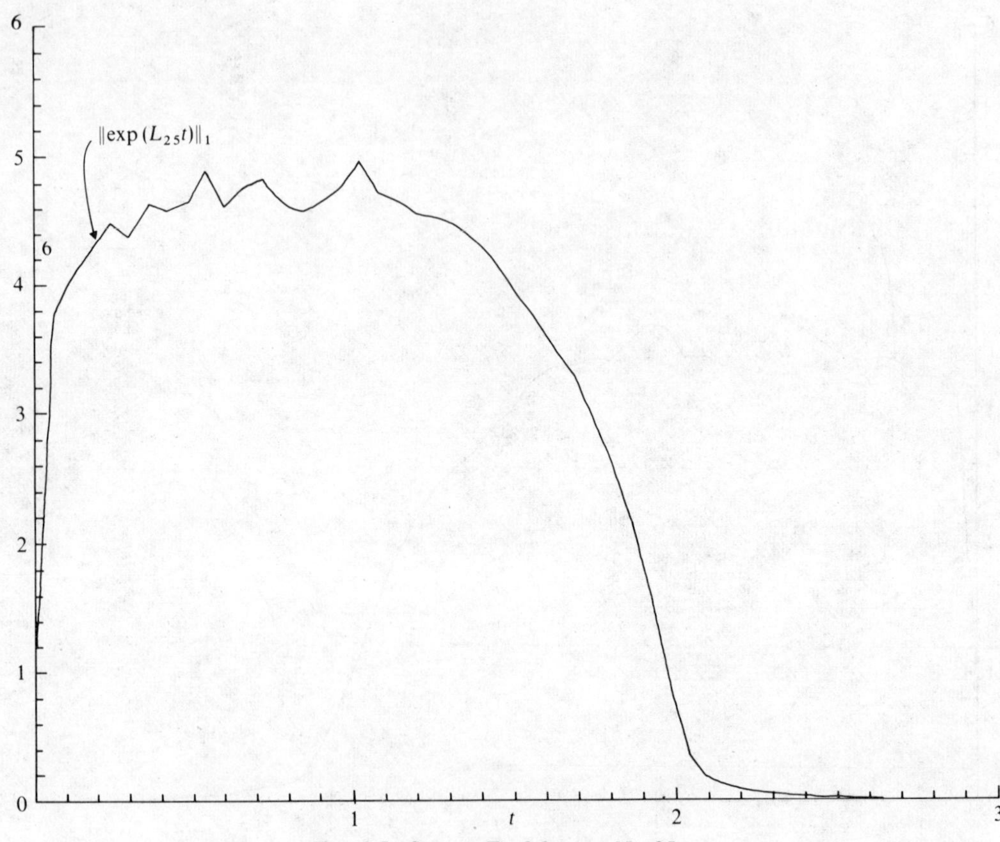

FIG. 8.5. *Same as Fig. 8.3 except $N = 25$.*

TABLE 8.4

The condition number $\|H_N\|\|H_N^{-1}\|$ in the L_2 matrix norm of the Lyapunov matrices H_N for the collocation, tau, and Galerkin spectral methods for (8.1)–(8.3). For the collocation and Galerkin methods, the condition number seems to grow at most like N^3 as $N \to \infty$, while for the tau method it seems to grow like N^2 as $N \to \infty$.

N	Collocation	Tau	Galerkin
10	4.1463×10^2	3.1090×10^2	4.6388×10^2
20	3.0332×10^3	1.2421×10^3	3.2672×10^3
30	9.8746×10^3	2.7938×10^3	1.0464×10^4
40	2.2940×10^4	4.9662×10^3	2.9083×10^4
50	4.4220×10^4	7.7593×10^3	4.6138×10^4

Let us now multiply both sides of (8.9) by $2(1-x)u_N$ and integrate with respect to the Chebyshev weight function $(1-x^2)^{-1/2}$. Thus, the left-hand side of (8.9) becomes

$$
\begin{aligned}
&2\int_{-1}^{1}(1-x)u_N\left[\frac{\partial u_N}{\partial t}+\frac{\partial u_N}{\partial x}\right](1-x^2)^{-1/2}\,dx\\
&\quad=\frac{d}{dt}\int_{-1}^{1}(1-x)(1-x^2)^{-1/2}u_N^2\,dx+\int_{-1}^{1}(1-x)^{1/2}(1+x)^{-1/2}\frac{\partial u_N^2}{\partial x}\,dx\\
&\quad=\frac{d}{dt}\int_{-1}^{1}(1-x)(1-x^2)^{-1/2}u_N^2\,dx+(1-x)^{1/2}(1+x)^{-1/2}u_N^2\Big|_{-1}^{1}\\
&\qquad+\int_{-1}^{1}u_N^2\left[\tfrac{1}{2}(1-x)^{-1/2}(1+x)^{-1/2}+\tfrac{1}{2}(1-x)^{1/2}(1+x)^{-3/2}\right]dx.
\end{aligned}
\tag{8.10}
$$

The boundary term in the last expression vanishes because u_N is a polynomial satisfying $u_N(-1)=0$. Also,

$$
\begin{aligned}
(1-x)u_N &= (1-x)\sum_{n=1}^{N}a_n[T_n-(-1)^nT_0]\\
&=\sum_{n=1}^{N}a_n[T_n-(-1)^nT_0]-\sum_{n=1}^{N}a_n[\tfrac{1}{2}(T_{n+1}+T_{n-1})-(-1)^nT_1]\\
&=\sum_{n=1}^{N}a_n[T_n-(-1)^nT_0]-\tfrac{1}{2}\sum_{n=1}^{N}a_n(T_{n+1}-(-1)^nT_1]\\
&\quad-\tfrac{1}{2}\sum_{n=1}^{N}a_n[T_{n-1}-(-1)^nT_1].
\end{aligned}
\tag{8.11}
$$

The first and third sums on the right in (8.11) are orthogonal to the right side of (8.9). The inner product of $(1-x)u_N$ with the second sum on the right in (8.9) gives

$$
-\frac{\pi}{4}(-1)^N\tau_Na_N=-\frac{\pi}{4}a_N\frac{da_N}{dt}.
\tag{8.12}
$$

Combining (8.10) and (8.12), we obtain

$$
\frac{1}{2}\frac{d}{dt}\int_{-1}^{1}(1-x)(1-x^2)^{-1/2}u_N^2\,dx+\frac{\pi}{8}\frac{d}{dt}a_N^2\leq 0.
\tag{8.13}
$$

This inequality proves that u_N is stable in the new norm defined in (8.13):

$$
\|u\|^2=\int_{-1}^{1}(1-x)(1-x^2)^{-1/2}|u(x)|^2\,dx.
\tag{8.14}
$$

Observe that (8.13) implies that the eigenvalues of L_N have nonpositive real parts.

It remains to prove that the norm defined by (8.14) is algebraically equivalent to the usual Chebyshev–L_2 norm. That is, we must show the existence of two functions $c_1(N)$ and $c_2(N)$ such that for every Nth degree polynomial u_N,

$$(8.15)\qquad c_1\int_{-1}^{1}\frac{u_N^2}{\sqrt{1-x^2}}\,dx \leqq \int_{-1}^{1}\frac{(1-x)u_N^2}{\sqrt{1-x^2}}\,dx \leqq c_2\int_{-1}^{1}\frac{u_N^2}{\sqrt{1-x^2}}\,dx$$

where $1/c_1(N)$ and $c_2(N)$ grow at most algebraically as $N\to\infty$. The second inequality in (8.15) holds with $c_2(N)=2$ because $1-x\leqq 2$.

The first inequality in (8.15) is more difficult to establish. By the mean value theorem,

$$\int_{-1}^{1}\frac{1-x}{\sqrt{1-x^2}}u_N^2\,dx=(1-\xi_N)\int_{-1}^{1}\frac{u_N^2}{\sqrt{1-x^2}}\,dx,\qquad -1<\xi_N<1.$$

However this does not prove the required inequality because it is not clear that $1/(1-\xi_N)$ is bounded algebraically as $N\to\infty$ for all polynomials.

To establish the first inequality in (8.15) we use a different approach. We substitute the Chebyshev polynomial expansion

$$u_N=\sum_{n=0}^{N}a_nT_n$$

and obtain

$$\begin{aligned}\frac{2}{\pi}\int_{-1}^{1}\frac{(1-x)u_N^2}{\sqrt{1-x^2}}\,dx&=2a_0^2-2a_0a_1+\sum_{n=1}^{N}a_n^2-\tfrac{1}{2}\sum_{n=2}^{N}(a_na_{n-1}+a_na_{n+1})\\&=(a_0\cdots a_N)H_N(a_0\cdots a_N)^T,\end{aligned}$$

where H_N is the symmetric, positive definite, $(N+1)\times(N+1)$ tridiagonal matrix whose elements are

$$(8.16)\qquad (H_N)_{jk}=\begin{cases}\tfrac{1}{2}c_j & \text{if } j=k,\\ -\tfrac{1}{2}c_j & \text{if } j=k-1,\\ -\tfrac{1}{2}c_k & \text{if } j=k+1,\\ 0 & \text{otherwise,}\end{cases}$$

where $c_0=2$, $c_n=1$ if $n>0$. To complete the demonstration of the first inequality in (8.15), we must show that $H_N\geqq c_1(N)I$ where $c_1(N)>0$ and $1/(c_1(N))$ is bounded algebraically as $N\to\infty$.

Since H_N is nearly a constant-diagonal tridiagonal matrix, the eigenvalues of H_N can be studied by standard techniques: if $D_N=\det(H_N-\lambda I)$, then D_N satisfies the three-term recurrence relation

$$(8.17)\qquad D_N=(1-\lambda)D_{N-1}-\tfrac{1}{4}D_{N-2},\qquad N\geqq 2.$$

Since (8.17) has constant coefficients, it is easy to solve exactly. From this solution, it is not hard to show that the smallest eigenvalue of H_N satisfies

$$\lambda_{\min}^{(N)} \sim \frac{\pi^2}{8N^2}, \qquad N\to\infty.$$

Choosing $c_1(N)=\lambda_{\min}(N)$ gives $1/(c_1(N))\sim 8N^2/\pi^2$ $(N\to\infty)$. This proves that the norm defined by (8.14) is algebraically equivalent to the Chebyshev norm and, therefore, Chebyshev–Galerkin approximation to (8.1) is algebraically stable. Note also that (8.13) shows that the matrix H_N defined in (8.16) satisfies (5.7b) with $c(N)=0$. Since $\|H_N\|=O(1)$ and $\|H_N^{-1}\|=O(N^2)$, then (5.11) implies that $\|\exp(L_N t)\|=O(N)$ as $N\to\infty$, which also follows directly from (8.15).

We have not yet been able to obtain a rigorous demonstration that $\|\exp(L_N t)\|=O(N^{1/2})$ as $N\to\infty$ as found numerically in Table 8.2. Our best result to date is $\|\exp(L_N t)\|=O(N)$ as $N\to\infty$.

Although the problem (8.1) is not well posed in the Chebyshev norm (as shown in § 5), it is well posed in the norm defined by (8.14). Using (8.1) and (8.3), we obtain

$$\int_{-1}^{1}\left(\frac{1-x}{1+x}\right)^{1/2} uu_t\,dx = -\int_{-1}^{1}\left(\frac{1-x}{1+x}\right)^{1/2} uu_x\,dx$$
$$= -\frac{1}{2}\int_{-1}^{1} u^2(1-x)^{-1/2}(1+x)^{-3/2}\,dx \leqq 0.$$

Thus,

$$\frac{d}{dt}\int_{-1}^{1}\left(\frac{1-x}{1+x}\right)^{1/2} u^2\,dx \leqq 0,$$

so that $\|e^{Lt}\|\leqq 1$ in the norm (8.14).

Proof of algebraic stability for Chebyshev–tau approximation. The proof of algebraic stability for the tau method is similar to that just given for Galerkin approximation. The Chebyshev–tau approximation u_N satisfies

$$\frac{\partial u_N}{\partial t}+\frac{\partial u_N}{\partial x} = \tau_N(t)T_N(x), \qquad u_N(-1,t)=0, \tag{8.18}$$

where

$$u_N=\sum_{n=0}^{N} a_n T_n. \tag{8.19}$$

Therefore,

$$(1-x)\frac{\partial^2 u_N}{\partial x\,\partial t} = -N\frac{da_N}{dt}T_N+\sum_{n=0}^{N-1} b_n T_n. \tag{8.20}$$

Moreover, comparing the coefficients of x^N on both sides of (8.18) we find

$$\tau_N(t)=\frac{da_N}{dt}. \tag{8.21}$$

Equations (8.18)–(8.21) imply

$$\left(\frac{\partial u_N}{\partial t},(1-x)\frac{\partial^2 u_N}{\partial x\,\partial t}\right)+\left(\frac{\partial u_N}{\partial x},(1-x)\frac{\partial^2 u_N}{\partial x\,\partial t}\right)=-\frac{\pi}{2}N\left(\frac{da_N}{dt}\right)^2. \tag{8.22}$$

Since

$$\left.\frac{\partial u_N}{\partial t}\right|_{x=-1}=0,$$

we obtain

$$2\left(\frac{\partial u_N}{\partial t},(1-x)\frac{\partial^2 u_N}{\partial x\,\partial t}\right)=\int_{-1}^{1}(1-x)^{1/2}(1+x)^{-1/2}\partial(\partial u_N/\partial t)^2/\partial x\,dx$$

$$=\int_{-1}^{1}(1-x)^{-1/2}(1+x)^{-3/2}(\partial u_N/\partial t)^2\,dx.$$

Therefore, (8.21) gives

$$\frac{d}{dt}\int_{-1}^{1}(1-x)(1-x^2)^{-1/2}\left(\frac{\partial u_N}{\partial x}\right)^2dx\leqq 0. \tag{8.23}$$

This proves that the evolution of $\partial u_N/\partial x$ is stable in the norm (8.14). Finally, the boundedness of $\partial u_N/\partial x$ implies the boundedness of u_N, as will now be shown. If u_N is given by (8.19), then

$$\frac{\partial u_N}{\partial x}=\sum_{n=0}^{N-1}b_nT_n$$

where

$$a_n=\frac{c_{n-1}b_{n-1}-b_{n+1}}{2n},\qquad n=1,\cdots,N.$$

The boundary condition $u_N(-1,t)=0$ requires that

$$a_0=\sum_{n=1}^{N}(-1)^{n+1}a_n.$$

Therefore, since $\partial u_N/\partial x$ is bounded algebraically as $N\to\infty$, so is u_N.

In § 11 we present a variety of numerical results for the numerical solution of (8.1) by Chebyshev and Legendre spectral methods.

Effect of boundary conditions on the stability of spectral methods. Let us discuss the effect of boundary conditions on the stability of the Chebyshev approximations to (8.1). In § 6 it was shown that incorrect treatment of the boundary does not affect the stability (though it does affect the convergence) of the Fourier–Galerkin method. This is not the case for the Chebyshev–spectral methods. Let us assume that we solve (8.1) ignoring the boundary condition (8.3) and suppose that $u_N(x, 0) = T_N(x)$. The resulting system of Galerkin equations for $\{a_n\}$ is

$$\frac{\partial a_n}{\partial t} = -\frac{2}{c_n} \sum_{\substack{p=n+1 \\ p+n \text{ odd}}}^{N} p a_p \tag{8.24}$$

where $a_n(0) = \delta_{nN}$. Equation (8.24) can easily be solved: $a_{N-k}(t)$ is a polynomial in t of degree k of the form

$$a_{N-k}(t) = \frac{1}{c_{N-k}}(-2)^k \binom{N}{k} t^k + \cdots. \tag{8.25}$$

This solution is clearly not bounded by any finite power of N. Thus, the Chebyshev methods are algebraically unstable when no boundary conditions are applied.

If we had imposed the boundary condition $u(+1, t) = 0$ in addition to, or instead of, the boundary condition $u(-1, t) = 0$, then Chebyshev–spectral solution to (8.1) would be unstable. With $u(+1, t) = 0$ instead of (8.3), the Chebyshev–spectral approximations to the operator $-\partial/\partial x$ all have eigenvalues with positive real parts (that grow as $N \to \infty$). Similarly, if we tried to impose the extra boundary condition $\partial u(+1, t)/\partial x = 0$ in addition to $u(-1, t) = 0$ (as is frequently done with finite difference methods), an unstable scheme would result.

The effect of imposing $u(+1, t) = 0$ in addition to $u(-1, t) = 0$ is slightly different for Legendre–spectral methods. With $u(-1, t) = u(+1, t) = 0$, Legendre–spectral methods for solution of (8.1) are semi-bounded. In fact,

$$(v, [L + L^*]v) = -2 \int_{-1}^{1} v \, \partial v/\partial x \, dx = 0$$

when $v(\pm 1, t) = 0$, so these methods are semi-bounded and stable. However, these spectral approximations are not consistent. For example, Galerkin approximation involves expansion of $u(x, t)$ in terms of the functions $\phi_{2n}(x) = P_{2n}(x) - P_0(x)$, $\phi_{2n+1}(x) = P_{2n+1}(x) - P_1(x)$ that satisfy $\phi_n(\pm 1) = 0$. But $\partial u/\partial x$ cannot, in general, be expanded in terms of the functions $\phi_n'(x)$.

The above situations are typical of rapidly converging spectral methods. Spectral methods are extremely sensitive to the proper formulation of boundary conditions. When proper boundary conditions are imposed so the problem is well posed, the methods yield very accurate results; when improper boundary conditions are mistakenly applied, the methods are likely to be explosively unstable. The formulation of stable and convergent spectral methods is strikingly similar to the formulation of well posed initial-value problems for the continuum equations.

SECTION 9

Time Differencing

In previous sections we have investigated the properties of spectral approximations to the spatial operator L of the differential equation

$$\frac{\partial u}{\partial t} = Lu.$$

In this section we investigate the properties of time-integration techniques for the solution of the semi-discrete spectral approximations

$$\frac{\partial u_N}{\partial t} = L_N u_N. \tag{9.1}$$

Time discretization errors in both finite-difference and spectral methods are typically much smaller than are spatial discretization errors. There are two reasons for this: (i) time steps are frequently restricted in size by explicit stability conditions—stability of the time integration *requires* that time-differencing errors be small; and (ii) many problems involve several space coordinates so any possible efficiency in the representation of the spatial variation of the dependent variables is quite important to the overall efficiency of the method—if the number of degrees of freedom necessary to describe a certain three-dimensional field accurately can be reduced by two in each space direction then the total number of degrees of freedom is decreased by a factor 8, but a similar improvement in time differencing gives just a factor 2. We will investigate here only finite-difference methods of finite-order accuracy for timewise solution of (9.1) despite the infinite-order accuracy in space of many of the spectral methods discussed in earlier sections. No efficient, infinite-order accurate time-differencing methods for variable coefficient problems are yet known. The current state-of-the-art of time-integration techniques for spectral methods is far from satisfactory on both theoretical and practical grounds and the results to be presented here must be regarded as only a beginning.

One of our prime goals is to investigate the *stability* of time-differencing methods for the solution of (9.1). To do this we must first explain how to extend the stability definitions given in §§ 4 and 5. Let $u_N^n(x) = \hat{u}_N(x, n\,\Delta t)$ be the approximation to the solution of (9.1) at time $n\,\Delta t$, where Δt is a time step. Time differencing methods involve approximating (9.1) in some way to give a rule for constructing u_N^{n+1}.

$$u_N^{n+1} = K_N(\Delta t) u_N^n, \tag{9.2}$$

where K_N is an operator acting on u_N. Using this rule repetitively it follows that

$$\hat{u}_N(x, n\,\Delta t) = [K_N(\Delta t)]^n u_N(x, 0), \tag{9.3}$$

where, for notational simplicity, we assume Δt fixed. We say that (10.2) is *strongly* stable if

$$\|[K_N(\Delta t)]^n\| \leqq K(n\,\Delta t) \tag{9.4}$$

for all N and n sufficiently large and Δt sufficiently small. Here $K(T)$ is a finite function of T. We define generalized stability by replacing $K(T)$ in (9.4) by $N^{r+sT}K(T)$ as in (5.2).

A sufficient, though not necessary, condition for strong stability (9.4) is

$$\|K_N(\Delta t)\| - 1 \leqq \kappa\,\Delta t \tag{9.5}$$

for some finite κ and all Δt sufficiently small. If $K_N(\Delta t)$ is a normal matrix then strong stability is assured in the L_2 matrix norm if the eigenvalues λ of K_N satisfy the von Neumann condition

$$\max|\lambda| \leqq 1 + \kappa\,\Delta t \tag{9.6}$$

for sufficiently small Δt. If K_N is not a normal matrix, then (9.6) is still a necessary, though not sufficient, condition for stability in the sense of (9.4).

The importance of these stability definitions is that they lead to the fully discrete form of the equivalence theorem (see § 4): a scheme is consistent if

$$\left\|\left(\frac{K_N(\Delta t) - I}{\Delta t} - L\right)u\right\| \to 0 \tag{9.7}$$

as $N \to \infty$ and $\Delta t \to 0$ for all u in a dense subspace of H; a scheme is convergent if

$$\|u_N^n - u(n\,\Delta t)\| \to 0$$

as $N \to \infty$ and $\Delta t \to 0$ for all n satisfying $0 \leqq n\,\Delta t \leqq T$ and all $u(0) \in \mathcal{H}$. The equivalence theorem states that for consistent approximations to well posed linear problems, stability is equivalent to convergence.

Let us now study the stability properties of some specific time-differencing methods.

Implicit time-integration methods. Two time-integration methods that are unconditionally stable in the generalized sense for every algebraically stable spectral method are the Crank–Nicolson scheme and the backwards Euler scheme. For any semi-discrete spectral approximation (9.1) to $u_t = Lu$, the Crank–Nicolson time-differencing scheme is given by

$$u_N^{n+1} - u_N^n = \Delta t\, L_N\left(\frac{u_N^{n+1} + u_N^n}{2}\right) \tag{9.8}$$

and the backwards Euler scheme is given by

$$u_N^{n+1} - u_N^n = \Delta t\, L_N u_N^{n+1}. \tag{9.9}$$

Let us prove that (9.8) and (9.9) are stable in the generalized sense. If (9.1) is algebraically stable there exists a family of positive definite Hermitian matrices $\{H_N\}$ such that

$$H_N L_N + L_N^* H_N \leqq \alpha(N) H_N$$

or, equivalently,

$$H_N^{1/2} L_N H_N^{-1/2} + H_N^{-1/2} L_N^* H_N^{1/2} \leqq \alpha(N) I,$$

where $\alpha(N) \leqq d \ln N$ for some finite d. Substituting

$$v_N^n = H_N^{1/2} u_N^n$$

into (9.8)–(9.9), we obtain, respectively,

$$v_N^{n+1} - v_N^n = \Delta t\, M_N \left(\frac{v_N^{n+1} + v_N^{n-1}}{2} \right), \tag{9.10}$$

$$v_N^{n+1} - v_N^n = \Delta t\, M_N v_N^{n+1}, \tag{9.11}$$

where

$$M_N = H_N^{1/2} L_N H_N^{-1/2}.$$

Taking the scalar product of (9.10) with $v_N^n + v_N^{n+1}$, we get

$$\begin{aligned} \|v_N^{n+1}\|^2 - \|v_N^n\|^2 &= \frac{\Delta t}{2} \left((v_N^{n+1} + v_N^n), \left(\frac{M_N + M_N^*}{2} \right) (v_N^{n+1} + v_N^n) \right) \\ &\leqq \frac{\alpha\, \Delta t}{4} \|v_N^{n+1} + v_N^n\|^2 \leqq \frac{\alpha\, \Delta t}{2} [\|v_N^{n+1}\|^2 + \|v_N^n\|^2]. \end{aligned} \tag{9.12}$$

Therefore, if $\alpha\, \Delta t < 2$,

$$\|v_N^{n+1}\|^2 \leqq \frac{(1 + \frac{1}{2}\alpha\, \Delta t)}{(1 - \frac{1}{2}\alpha\, \Delta t)} \|v_N^n\|^2, \tag{9.13}$$

which proves generalized stability for v_N and, hence, also for

$$u_N = H_N^{-1/2} v_N.$$

Similarly, we may show that the backwards Euler method is unconditionally stable in the generalized sense. Taking the scalar product of (9.11) with $v_N^{n+1} + v_N^n$, we obtain

$$\begin{aligned} \|v_N^{n+1}\|^2 - \|v_N^n\|^2 &= \Delta t\, (M_N v_N^{n+1}, v_N^n + v_N^{n+1}) \\ &= \Delta t\, (M_N v_N^{n+1}, 2 v_N^{n+1} - \Delta t\, M_N v_N^{n+1}) \\ &\leqq \alpha\, \Delta t\, \|v_N^{n+1}\|^2, \end{aligned} \tag{9.14}$$

so that, if $\alpha\, \Delta t < 1$,

$$\|v_N^{n+1}\|^2 \leqq \frac{1}{1 - \alpha\, \Delta t} \|v_N^n\|^2,$$

proving generalized stability of v_N and, hence, u_N.

Note that the above proofs show that if $\alpha(N)$ is a bounded function of N then $v_N = H_N^{1/2} u_N$ is strongly stable for both the Crank–Nicolson and backwards Euler schemes.

Spectral approximations using Fourier series. Next, we consider several time integration methods for Fourier series spectral approximations to

$$u_t + u_x = 0$$

with periodic boundary conditions. As shown in § 6, the collocation equations are

$$\frac{\partial u_N}{\partial t} = C^{-1} D C u_N \tag{9.15}$$

where the $2N \times 2N$ matrices C and D are defined in (6.3).

The "leapfrog" time differencing approximation to (9.15) is the explicit two-level scheme

$$u_N^{n+1} - u_N^{n-1} = 2\,\Delta t\, C^{-1} D C u_N^n. \tag{9.16}$$

Thus, in the leapfrog scheme

$$K_N(\Delta t) u_N^n = u_N^{n-1} + 2\,\Delta t\, C^{-1} D C u_N^n,$$

so K_N is a two-level evolution operator since it depends on both u_N^{n-1} and u_N^n. The definitions of stability, convergence, and consistency given above extend easily to this case.

We shall show that (9.16) is strongly stable provided that

$$\Delta t < \frac{1}{2\pi(N-1)}. \tag{9.17}$$

To show this we first recall from § 6 that C is unitary and D is skew-Hermitian. Therefore, $A = C^{-1} D C$ is also skew-Hermitian, and hence normal, so that

$$\|A\| = 2\pi(N-1).$$

Now we take the inner product of (9.16) with $u_N^{n+1} + u_N^{n-1}$ to get

$$\|u_N^{n+1}\|^2 - \|u_N^{n-1}\|^2 = 2\,\Delta t\,\mathrm{Re}\,(u_N^{n+1} + u_N^{n-1}, A u_N^n),$$

since $u_N^{n\pm1}$ and u_N^n are real. Since $A^* = -A$, we obtain

$$\begin{aligned} U_N^n &\equiv \|u_N^{n+1}\|^2 + \|u_N^n\|^2 - 2\,\Delta t\,\mathrm{Re}\,(u_N^{n+1}, A u_N^n) \\ &= \|u_N^n\|^2 + \|u_N^{n-1}\|^2 - 2\,\Delta t\,\mathrm{Re}\,(u_N^n, A u_N^{n-1}) \equiv U_N^{n-1} \end{aligned}$$

so $U_N^n = U_N^0$. Schwarz' inequality implies that

$$|\mathrm{Re}\,(u_N^{n+1}, A u_N^n)| \leqq \|A\|\,\|u_N^{n+1}\|\,\|u_N^n\|$$

so that if (9.17) is satisfied, i.e., $\Delta t \|A\| \leqq 1 - \varepsilon$ for some $\varepsilon > 0$, then

$$|2\,\Delta t\,\mathrm{Re}\,(u_N^{n+1}, A u_N^n)| \leqq 2(1-\varepsilon)\|u_N^{n+1}\|\,\|u_N^n\|.$$

Using this result, we obtain

$$\varepsilon(\|u_N^{n+1}\|^2+\|u_N^n\|^2)+(1-\varepsilon)(\|u_N^{n+1}\|-\|u_N^n\|)^2 \leqq U_N^n = U_N^0. \tag{9.18}$$

Since U_N^0 is a bounded function of N (because of the smoothness of the initial conditions), we see that $\|u_N^{n+1}\|$ is bounded for all N and n, proving strong stability.

Another way to prove that the leapfrog and Crank–Nicolson time differencing schemes are strongly stable for (9.15) is to use a modal analysis, which is justified because A is normal. Thus, if u_N^0 is an eigenfunction of A with *eigenvalue* λ, the Crank–Nicolson approximation to $K_N(\Delta t)$ is

$$K_N(\Delta t)u_N^0 = (1+\tfrac{1}{2}\lambda\,\Delta t)/(1-\tfrac{1}{2}\lambda\,\Delta t)u_N^0. \tag{9.19}$$

Since the eigenvalues λ of $C^{-1}DC$ are all pure imaginary, it follows that $\|K_N(\Delta t)\|=1$, so Crank–Nicolson differencing is stable.

Still another time-differencing method for solution of (9.15) is to use a Runge–Kutta scheme. It turns out that the first- and second-order Runge–Kutta methods are unstable unless Δt satisfies conditions that are much more restrictive than (9.17). With the first-order Euler method

$$u_N^{n+1} = u_N^n + \Delta t\, A u_N^n,$$

stability requires that $N^2\,\Delta t$ be bounded as $\Delta t \to 0$ (because $\|K_N(\Delta t)\| = 1+O(N^2\,\Delta t^2)$); with the second-order scheme

$$\tilde{u}_N^{n+1/2} = u_N^n + \tfrac{1}{2}\Delta t\, A u_N^n,$$

$$u_N^{n+1} = u_N^n + \Delta t\, A\tilde{u}_N^{n+1/2},$$

stability requires that $N^{4/3}\,\Delta t$ be bounded as $\Delta t \to 0$. However, the third and fourth-order Runge–Kutta methods give conditional stability restrictions like (9.17) which we will now derive.

The third-order Runge–Kutta scheme may be written for a linear equation like (9.1) as

$$u_N^{n+1} = [I+\Delta t\, A+\tfrac{1}{2}(\Delta t\, A)^2+\tfrac{1}{6}(\Delta t\, A)^3]u_N^n = K_N(\Delta t)u_N^n. \tag{9.20}$$

Since $K_N(\Delta t)$ given by (9.20) is normal,

$$\|K_N(\Delta t)\| = \max_\lambda |1+\lambda\,\Delta t+\tfrac{1}{2}(\lambda\,\Delta t)^2+\tfrac{1}{6}(\lambda\,\Delta t)^3|,$$

where the maximum is taken over all the eigenvalues of A. Eigenvalues of A are ik with $|k| \leqq 2\pi(N-1)$, so (9.6) is satisfied provided that

$$\Delta t < \frac{\sqrt{3}}{2\pi(N-1)}. \tag{9.21}$$

Thus, this method allows time steps that can be $\sqrt{3}$ times larger than with the leapfrog scheme while maintaining stability. However, if the operator A is complicated, the third-order Runge–Kutta scheme requires about 3 times as much work as leapfrog at each time step, so it is probably not competitive.

Similar analysis of the fourth-order Runge–Kutta scheme gives the stability condition

$$\Delta t < \frac{\sqrt{2}}{\pi(N-1)}. \tag{9.22}$$

Thus time steps can be nearly three times larger than with leapfrog steps. However, fourth-order Runge–Kutta differencing requires about four times the work of leapfrog differencing, so the scheme is probably not very useful unless high accuracy is desired.

Now we shall consider time-differencing methods for Fourier series spectral approximations to the heat equation with periodic boundary conditions:

$$u_t = u_{xx}, \qquad 0 \leqq x \leqq 1. \tag{9.23}$$

Collocation using Fourier series gives the spectral equations

$$\frac{\partial u_N}{\partial t} = C^{-1}D^2Cu_N. \tag{9.24}$$

The matrix $C^{-1}D^2C$ is negative definite. Because (9.19) still holds and all eigenvalues λ are negative, Crank–Nicolson time differencing is unconditionally stable. On the other hand, it is easy to show that leapfrog differencing is unconditionally unstable. In fact, if u_N^0 is an eigenfunction of $C^{-1}D^2C$ with eigenvalue $\lambda < 0$, then $\|K_N(\Delta t)^n u_N^0\|$ grows like $(-\lambda\,\Delta t + \sqrt{1+(\lambda\,\Delta t)^2})^n \sim e^{-\lambda(n\,\Delta t)}$ as $\Delta t \to 0$ for fixed λ and $n\,\Delta t$. Since $\max|\lambda| = 4\pi^2(N-1)^2$ grows like N^2 as $N\to\infty$, $\|K_N(\Delta t)^n u_N^0\|$ cannot be bounded by a finite function of $n\,\Delta t$ for all N, proving unconditional instability.

Another way to solve (9.24) is to use a generalized Dufort–Frankel scheme

$$\frac{u_N^{n+1}-u_N^{n-1}}{2\,\Delta t} = C^{-1}D^2Cu_N^n - \gamma N^2(u_N^{n+1} - 2u_N^n + u_N^{n-1}). \tag{9.25}$$

If $\gamma \geqq \pi^2$, then this method is unconditionally stable (Gottlieb and Gustaffson (1976)).

Similarly, Euler time differencing of (9.24) is conditionally stable. Stability requires that

$$\Delta t \max|\lambda| \leqq 2 \quad \text{or} \quad \Delta t \leqq [2\pi^2(N-1)^2]^{-1}. \tag{9.26}$$

Higher-order Adams–Bashforth schemes have similar conditional stability limits.

Time-differencing for mixed initial-boundary value problems. Some care is necessary in the formulation of time-differencing methods for spectral approximations to mixed initial-boundary value problems. The sensitivity of spectral methods to the proper formulation of boundary conditions, as shown in §§ 6–8, carries over to the formulation of time-differencing methods for these approximations. For example, for most mixed initial-boundary value problems leapfrog time differencing is unconditionally unstable for spectral approximations. Furthermore, explicit time integration methods may be unduly restricted by conditional stability requirements in spectral approximations. The origin of these severe restrictions is the very high resolution of spectral methods near boundaries. Thus, it is frequently necessary to combine special kinds of implicit time-integration methods with spectral approximations in order to maintain high accuracy at reasonable computational cost. Several examples will be given later.

Let us begin by studying time-differencing methods for the Chebyshev–spectral approximation to the mixed initial-boundary value problem (8.1)–(8.3);

$$u_t + u_x = 0, \qquad -1 \leqq x \leqq 1, \quad t > 0, \tag{9.27}$$

$$u(x, 0) = f(x), \qquad -1 \leqq x \leqq 1, \tag{9.28}$$

$$u(-1, t) = 0, \qquad t > 0. \tag{9.29}$$

In § 8, we proved that various semi-discrete spectral approximations to (9.27)–(9.29) are algebraically stable.

Let us first consider the leapfrog time-differencing scheme

$$u_N^{n+1} = u_N^{n-1} + 2\,\Delta t L_N u_N^n, \tag{9.30}$$

where $u_N^n(x)$ is the time-discretized approximation to $u_N(x, n\,\Delta t)$, Δt is the time step, and the semi-discrete approximation is $\partial u_N/\partial t = L_N u_N$.

This scheme is unconditionally unstable for any Δt as $N \to \infty$. To show this we recall that in § 8, we proved that the eigenvalues of L_N have negative real part (see Table 8.3) and that the largest eigenvalue of L_N has a negative real part that grows like N^2 as $N \to \infty$. Let us rewrite (9.30) in the 2×2 block-matrix form

$$\begin{pmatrix} u_N^{n+1} \\ u_N^n \end{pmatrix} = \begin{pmatrix} 2\,\Delta t\, L_N & I \\ I & 0 \end{pmatrix} \begin{pmatrix} u_N^n \\ u_N^{n-1} \end{pmatrix}. \tag{9.31}$$

If the eigenvalues of L_N are denoted as μ_N, then the eigenvalues of the matrix on the right in (9.31) are

$$\lambda_N^{(\pm)} = \mu_N\,\Delta t \pm \sqrt{1 + (\Delta t)^2 \mu_N^2}. \tag{9.32}$$

For fixed N and $\Delta t \to 0$,

$$\lambda_N^{(-)} = e^{-\mu_N \Delta t}(-1 + O(\Delta t^2)). \tag{9.33}$$

Thus

$$[\lambda_N^{(-)}]^n = (-1)^n e^{-\mu_N n\,\Delta t}(1 + O(\Delta t)), \quad 0 \leqq n\,\Delta t \leqq T, \quad \Delta t \to 0. \tag{9.34}$$

Since $\|K_N(\Delta t)^n\| \geqq |\lambda_N^{(-)}|^n$ and there are eigenvalues of L_N with negative real part of order N^2, no inequality of the form (9.4) can be satisfied. Thus, leapfrog time differencing of the Chebyshev approximations to (9.27)–(9.29) is unconditionally unstable.

There are several conditionally stable explicit time-differencing approximations that can be used with spectral approximations to (9.27)–(9.29). Two examples are the Adams–Bashforth scheme

$$u_N^{n+1}=u_N^n+\tfrac{3}{2}\Delta t\,L_N^n-\tfrac{1}{2}\Delta t\,L_N^{n-1} \tag{9.35}$$

and the modified Euler scheme

$$\hat{u}_N^{n+1}=u_N^n+\Delta t\,L_N u_N^n, \tag{9.36a}$$

$$u_N^{n+1}=u_N^n+\tfrac{1}{2}\Delta t\,L_N u_N^n+\tfrac{1}{2}\Delta t\,L_N\hat{u}_N^{n+1}. \tag{9.36b}$$

The modified Euler scheme (9.36) is in practice stable provided the stability condition

$$\Delta t\leqq 8/N^2 \tag{9.37}$$

is satisfied. A similar stability condition holds for the Adams–Bashforth scheme.

The fact that the stability limit in (9.37) depends on $1/N^2$ rather than $1/N$ is not very surprising because the Chebyshev collocation points $\{\cos \pi n/N;\ n=0, 1, \cdots, N\}$ are spaced by a distance of order $1/N^2$ near the boundaries. Since the wave speed in (9.27) is 1 the wave propagates from one grid point to the next in a time of order $1/N^2$ so time steps must be smaller than this to maintain explicit stability.

The explicit stability restriction (9.37) for Chebyshev–spectral methods with N polynomials should be contrasted with the corresponding stability conditions for finite difference approximations to (9.27)–(9.29). With N gridpoints uniformly spaced in the interval $-1\leqq x\leqq 1$, the grid spacing is $2/N$ so the Courant stability condition is $\Delta t\leqq 2/N$. As $N\to\infty$, this stability condition on finite difference schemes is much weaker than the condition (9.37) on the spectral approximations. A semi-implicit technique that permits stable time-differencing with spectral methods with a stability condition like that of finite-difference schemes will be discussed later in this section.

In order to prove that the modified Euler method (9.36) is stable, we begin by noting that (9.36) is equivalent to the second-order Taylor series method

$$u_N^{n+1}=(I+\Delta t\,L_N+\tfrac{1}{2}(\Delta t)^2L_N^2)u_N^n\equiv G_N u_N^n. \tag{9.38}$$

A sufficient condition for algebraic stability of (9.38) is the existence of positive-definite symmetric matrices S_N such that

$$G_N^T S_N G_N\leqq S_N \tag{9.39a}$$

and the condition number of S_N satisfies

$$\|S_N\|\|S_N^{-1}\|=O(N^\beta),\qquad N\to\infty. \tag{9.39b}$$

for some finite β. If (9.39) holds then

$$(G_N^T)^n S_N(G_N)^n\leqq (G_N^T)^{n-1}S_N(G_N)^{n-1}\leqq\cdots\leqq S_N$$

so that

$$S_N^{-1/2}(G_N^T)^n S_N^{1/2} S_N^{1/2}(G_N)^n S_N^{-1/2} \leqq I.$$

Therefore,

$$\|S_N^{1/2}(G_N)^n S_N^{-1/2}\| \leqq 1,$$

so that

$$\|u_N^n\| = \|(G_N)^n u_N^0\| \leqq \|S_N^{-1/2}\| \|S_N^{1/2}(G_N)^n S_N^{-1/2}\| \cdot \|S_N^{1/2}\| \|u_N^0\| = O(N^\beta \|u_N^0\|), \qquad N \to \infty.$$

To complete the stability proof we must investigate under what conditions matrices S_N satisfying (9.39) exist. One choice for S_N is just the Lyapunov matrices of L_N; these matrices satisfy

$$S_N L_N + L_N^T S_N = -I. \tag{9.40}$$

From Table 8.4 we observe that the Lyapunov matrices for spectral approximations to (9.27)–(9.29) have algebraically bounded condition number. Using (9.38), we obtain

$$G_N^T S_N G_N = [I + \Delta t\, L_N^T + \tfrac{1}{2}(\Delta t)^2 (L_N^2)^T] S_N [I + \Delta t\, L_N + \tfrac{1}{2}(\Delta t)^2 (L_N)^2]$$

or

$$\begin{aligned} G_N^T S_N G_N = S_N &+ \Delta t\, (L_N^T S_N + S_N L_N) \\ &+ \tfrac{1}{2}(\Delta t)^2 [(L_N^2)^T S_N + 2 L_N^T S_N L_N + S_N L_N^2] \\ &+ \tfrac{1}{2}(\Delta t)^3 [(L_N^2)^T S_N L_N + L_N^T S_N L_N^2] + \tfrac{1}{4}(\Delta t)^4 (L_N^2)^T S_N L_N^2. \end{aligned}$$

From (9.40), it follows that

$$\begin{aligned} (L_N^2)^T S_N + L_N^T S_N L_N &= -L_N^T, \\ L_N^T S_N L_N + S_N L_N^2 &= -L_N, \\ (L_N^2)^T S_N L_N + L_N^T S_N L_N^2 &= -L_N^T L_N \end{aligned}$$

so that

$$\begin{aligned} G_N^T S_N G_N = S_N - \Delta t\, I &- \tfrac{1}{2}(\Delta t)^2 [L_N^T + L_N] \\ &- \tfrac{1}{2}(\Delta t)^3 L_N^T L_N + \tfrac{1}{4}(\Delta t)^4 (L_N^2)^T S_N L_N^2. \end{aligned}$$

Thus, (9.39a) is satisfied provided that

$$-\Delta t\, (L_N^T + L_N) \leqq 2I, \tag{9.41}$$

$$\Delta t\, L_N^T S_N L_N \leqq 2I. \tag{9.42}$$

If (9.41)–(9.42) are satisfied then the modified Euler method for (9.27)–(9.29) is algebraically stable.

At first, it may appear that the stability condition (9.42) is much more severe than the stability condition (9.41). The Lyapunov matrices of Chebyshev polynomial approximations to the wave equation satisfy $\|S_N\| \geqq O(1)$ as $N \to \infty$ while the operator L_N satisfies $\|L_N\| = O(N^2)$ [see § 8], so that (9.42) seems to require that $\Delta t = O(1/N^4)$ as $N \to \infty$. However, the stability condition (9.42) is no more restrictive than the stability condition (9.41). To see this we use (9.40) written in the form

$$L_N^T S_N L_N L_N^{-1} + (L_N^T)^{-1} L_N^T S_N L_N = -I$$

to obtain the representation [see (5.13)]

$$L_N^T S_N L_N = \int_0^\infty \exp\,[(L_N^{-1})^T t]\exp\,[L_N^{-1} t]\,dt. \tag{9.43}$$

It may be shown that the norm of the integrand of (9.43) is $O(1)$ as $N \to \infty$ for $t = O(N^2)$ and that the norm decays rapidly to zero as $t \to \infty$. Therefore,

$$\|L_N^T S_N L_N\| = O(N^2), \qquad N \to \infty, \tag{9.44}$$

showing that the stability condition (9.42) is of the form $\Delta t = O(1/N^2)$.

Semi-implicit methods. When explicit time-stepping methods are used to solve semi-discrete spectral equations for the hyperbolic problem

$$\frac{\partial u}{\partial t} + a(x)\frac{\partial u}{\partial x} = 0, \qquad -1 \leqq x \leqq 1, \tag{9.45}$$

with appropriate boundary conditions (that depend on the sign of $a(x)$), there result stability conditions of the form

$$\Delta t \lesssim \min\left\{\frac{1}{N^2|a(1)|}, \frac{1}{N^2|a(-1)|}, \frac{1}{N \max_{|x|\leqq 1}|a(x)|}\right\}. \tag{9.46}$$

These stability limits can be derived heuristically from the Courant stability condition

$$\Delta t < \Delta x_{\text{eff}}/|a_{\text{eff}}| \tag{9.47}$$

where a_{eff} is the effective wave propagation speed in a direction in which there is effective grid resolution Δx_{eff}. Near the boundaries $x = \pm 1$, the Chebyshev–spectral methods have resolution $\Delta x_{\text{eff}} = O(1/N^2)$ as $N \to \infty$ while $a_{\text{eff}} = a(\pm 1)$; in the interior of $-1 < x < 1$, Chebyshev series have effective resolution $\Delta x_{\text{eff}} = O(1/N)$ as $N \to \infty$ while the largest wave speed is $\max|a(x)|$. Thus, (9.47) implies (9.46) for the Chebyshev–spectral methods.

The stability condition (9.46) is too severe for many applications because it requires that $\Delta t = O(1/N^2)$. In order to relax this severe constraint, we use a semi-implicit method in which the propagation through the high-resolution boundary is treated implicitly, but the propagation through the interior is treated explicitly.

One possible semi-implicit scheme is the following two-step method. Let L_N be the Chebyshev–spectral approximation to $-a(x)\,\partial/\partial x$ with appropriate boundary conditions applied, and L_N^+, L_N^- be the Chebyshev–spectral approximations to the constant coefficient wave operators $-a(+1)\,\partial/\partial x$, $-a(-1)\,\partial/\partial x$, respectively, again with appropriate boundary conditions applied. A semi-implicit two-step scheme is given by

(9.48a) $$u_N^{n+1/2}-\tfrac{1}{2}\Delta t\,L_N^-u_N^{n+1/2}=u_N^n+\tfrac{1}{2}\Delta t\,(L_N-L_N^-)u_N^n,$$

(9.48b) $$u_N^{n+1}-\tfrac{1}{2}\Delta t\,L_N^+u_N^{n+1}=u_N^{n+1/2}+\tfrac{1}{2}\Delta t\,(L_N-L_N^+)u_N^{n+1/2}.$$

The scheme (9.48) is stable if the stability condition

(9.49) $$\Delta t \lesssim \frac{1}{N\max|a(x)|}$$

is satisfied.

The condition (9.49) is sufficient to ensure stability, but the semi-implicit scheme (9.48) may be stable even if (9.49) is violated. If $\max|a(x)|<|a(1)|$ or $\max|a(x)|<|a(-1)|$, (9.48) is usually unconditionally stable for sufficiently large N (see § 8 of Orszag (1974)). The implementation of (9.48) on a computer is straightforward and efficient; the properties of Chebyshev polynomials summarized in the Appendix show that the implicit equations (9.48) are essentially tridiagonal matrix equations.

The reason that the semi-implicit method outlined above does not have a stability restriction like $\Delta t=O(1/N^2)$ can be understood as follows. By subtracting L_N^+ and L_N^- in succeeding half time-steps, the explicit part of the calculation is similar to that in solving an equation of the form

(9.50) $$\frac{\partial u}{\partial t}+(1-x^2)b(x)\frac{\partial u}{\partial x}=0$$

where the wave speed vanishes at $x=\pm1$. If $b(x)=b$, a constant, the Chebyshev–tau equations for (9.50) are just

(9.51) $$\frac{da_n}{dt}=2\frac{1}{c_n}b[(n-1)a_{|n-1|}-(n+1)a_{n+1}]$$

where $c_0=2$ and $c_n=1$ for $n>0$. By Gerschgorin's theorem, $\|L_N\|$ for (9.51) satisfies

(9.52) $$\|L_N\|=O(bN),\qquad N\to\infty,$$

so the explicit time step restriction is $\Delta t=O(1/bN)$ as $N\to\infty$.

We note that Chebyshev–spectral approximations to (9.50) are stable when no boundary conditions are applied. In fact, using Galerkin approximation and the Chebyshev inner product, we obtain

$$\left(u_N,\frac{\partial u_N}{\partial t}+b(1-x^2)\frac{\partial u_N}{\partial x}\right)=0,$$

so

$$\frac{d}{dt}\int_{-1}^{1}\frac{u_N^2}{\sqrt{1-x^2}}\,dx=-b\int_{-1}^{1}\sqrt{1-x^2}\,\frac{\partial}{\partial x}u_N^2\,dx$$

$$=-b\int_{-1}^{1}\frac{xu_N^2}{\sqrt{1-x^2}}\,dx\le|b|\int_{-1}^{1}\frac{u_N^2}{\sqrt{1-x^2}}\,dx.$$

Therefore,

$$\|u_N(t)\|^2\le e^{|b|t}\|u_N(0)\|^2,$$

proving stability.

There are other attractive semi-implicit schemes for (9.45). For example, suppose $a(x)$ is one-signed, say $a(x)>0$, and let $a_{\max}=\max a(x)$. Define $L_N^{\max}$ as the Chebyshev approximation to $-\frac{1}{2}a_{\max}\,\partial/\partial x$ with boundary conditions imposed at $x=-1$. A semi-implicit Chebyshev spectral scheme for (9.45) is

$$u_N^{n+1}-\Delta t\,L_N^{\max}u_N^{n+1}=u_N^n+\Delta t\,(L_N-L_N^{\max})u_N^n. \tag{9.53}$$

The scheme (9.53) is usually unconditionally stable and avoids the severe time step restriction (9.46). It is also easy to implement efficiently because $L_N^{\max}$ is a Chebyshev approximation to a constant-coefficient wave operator.†

The same kind of trick stabilizes spectral methods for non-linear equations. For example, if we are solving the equation

$$\frac{\partial u}{\partial t}+u\frac{\partial u}{\partial x}=0$$

during a time interval in which $u(x,t)$ is smooth (no shock waves), then we may use the semi-implicit scheme

$$\frac{\partial u}{\partial t}+\frac{1}{2}u_{\max}\frac{\partial u}{\partial x}=\left(\frac{1}{2}u_{\max}-u\right)\frac{\partial u}{\partial x}$$

in which the terms on the left are treated implicitly in time, while those on the right are treated explicitly. Here $u_{\max}$ is an estimate of the largest value of $u(x,t)$. Similar semi-implicit methods are effective in eliminating (or at least relaxing) time-step restrictions for finite-difference methods. The key idea is to recognize the source term of a numerical instability and then to approximate it by a simple expression that can easily be treated implicitly.

Several other examples of semi-implicit methods should make the general technique clear. For the variable coefficient heat equation

$$u_t=k(x)u_{xx},\qquad -1\le x\le 1,$$

† D. Haidvogel has pointed out that the semi-implicit scheme (9.53) with $L_N^{\max}$ replaced by a Chebyshev spectral approximation to $\frac{1}{2}(bx+c)\,\partial/\partial x$, where $b+c=a(+1)$, $c-b=a(-1)$, is also stable under the weak restriction (9.49). The resulting implicit equations are still tridiagonal [see (A.9), (A.18)].

with suitable boundary conditions at $x=\pm 1$ and $k(x)>0$, Chebyshev–spectral methods give explicit time-step stability conditions of the form

$$\Delta t \leq \min\left\{\frac{1}{k(-1)N^4}, \frac{1}{k(1)N^4}, \frac{1}{N^2 \max_{|x|<1} k(x)}\right\}. \tag{9.54}$$

The very severe time step restriction that $\Delta t = O(1/N^4)$ as $N \to \infty$ is due to the high resolution of Chebyshev series near the boundaries $x=\pm 1$. To avoid this problem we can use a semi-implicit method. Let L_N be the Chebyshev–spectral approximation to $k(x)\partial^2/\partial x^2$ and let $L_N^{\max}$ be the Chebyshev–spectral approximation to $\frac{1}{2}k_{\max}\,\partial^2/\partial x^2$ where $k_{\max} = \max k(x)$. The semi-implicit scheme (9.53) with $L_N^{\max}$ defined in this way seems to be unconditionally stable (Orszag (1974)) and certainly does not have any stability restrictions of the form (9.54).

Finally, we comment on the need for implicit or semi-implicit schemes in multi-dimensional problems. If we wish to solve the Navier–Stokes equations

$$\begin{aligned} &\frac{\partial \mathbf{u}}{\partial t} + \mathbf{u}\cdot\nabla\mathbf{u} = -\nabla p + \nu\nabla^2\mathbf{u}, \\ &\nabla\cdot\mathbf{u} = 0 \end{aligned} \tag{9.55}$$

for incompressible fluid flow, the treatment of the various terms should be guided closely by the type of stability restrictions they impose.

If $\nu = 0$ then we need only consider the types of stability restrictions induced by the advective term $-\mathbf{u}\cdot\nabla\mathbf{u}$ and by the pressure term $-\nabla p$; we will not discuss the effect of the pressure because it is closely connected to the incompressibility condition $\nabla\cdot\mathbf{u}=0$ and is not relevant to the semi-implicit ideas discussed here. At a boundary of the flow, it is appropriate to specify boundary conditions on $\mathbf{u}\cdot\mathbf{n}$ where $\mathbf{n}$ is the normal to the boundary. If the boundary is solid and stationary, then $\mathbf{u}\cdot\mathbf{n}=0$ and we are in a situation similar to that modeled by (9.50). The effective convective speed normal to the boundary vanishes, so spectral methods exhibit no unusual time stepping restrictions. However, if fluid is being blown into or sucked out of the boundary so $\mathbf{u}\cdot\mathbf{n}\neq 0$, then semi-implicit methods must be applied to avoid unreasonably restrictive conditions like (9.46) on the time steps.

If $\nu > 0$, then the viscous terms in the Navier–Stokes equations should be treated implicitly to avoid unreasonable time step restrictions due to the high resolution of spectral approximations near the boundary.

SECTION 10

Efficient Implementation of Spectral Methods

There are three aspects of the efficient implementation of spectral methods that we discuss here: (i) evaluation of derivatives; (ii) evaluation of nonlinear and nonconstant coefficient terms; (iii) roundoff errors. More details on these matters are given elsewhere (see the References).

Evaluation of derivatives. An efficient procedure to obtain the expansion coefficients of derivatives of a function $f(x)$ in terms of the expansion coefficients of $f(x)$ is to use recurrence relations. For example, to evaluate the term

$$S_n = \sum_{\substack{p=n+1 \\ p+n \text{ odd}}}^{N} p a_p$$

that appears in the Chebyshev equations (2.12), (2.20), and (2.33), we use the recurrence

$$S_n = S_{n+2} + (n+1) a_{n+1}, \qquad 0 \leqq n \leqq N-1, \tag{10.1}$$

with $S_N = S_{N+1} = 0$. In this way, S_n is evaluated for all n using only N arithmetic operations. The existence of the recurrence relation (10.1) is ensured by the recurrence property

$$2T_n = \frac{T'_{n+1}}{n+1} - \frac{T'_{n-1}}{n-1}, \qquad n > 1,$$

satisfied by the Chebyshev polynomials. Similarly, it is possible to derive recurrence relations to evaluate efficiently the coefficients of arbitrary derivatives of functions expanded in Chebyshev and other classical polynomial expansions.

Evaluation of nonlinear and nonconstant coefficient terms. The most efficient way to evaluate nonlinear and general nonconstant terms in spectral approximations is to apply transform methods. The key idea is to apply fast Fourier transforms and other transforms to transform efficiently between spectral representations of a function *f(x)* and physical-space representations of *f(x)*. With Chebyshev polynomial expansions, fast Fourier transforms permit the evaluation of arbitrary nonlinear and nonconstant coefficients terms in order $N \log N$ arithmetic operations.

In general, collocation methods give approximations to nonlinear and nonconstant coefficient problems that can be more efficiently implemented than Galerkin or tau approximations. Collocation is recommended for these problems. For example, the solution of the hyperbolic problem

$$(10.2)\qquad \frac{\partial u}{\partial t}+e^{u+x}\frac{\partial u}{\partial x}=f(x,t), \qquad -1\leqq x\leqq 1,\quad t>0,$$
$$u(-1,t)=0,$$

would be difficult using Galerkin or tau approximation but is straightforward using collocation methods.

Let us explain how to march the solution to (10.2) forward by one time step efficiently using Chebyshev collocation. We introduce the $N+1$ collocation points $x_j=\cos \pi j/N$ $(j=0,\cdots,N)$ and represent the current solution u_j as

$$(10.3)\qquad u_j=\sum_{n=0}^{N} a_n \cos\frac{\pi nj}{N}.$$

Then we invert (10.3) by the fast Fourier transform to obtain a_n for $n=0,1,\cdots,N$ and calculate

$$a_n^{(1)}=2S_n/c_n$$

by (10.1). Next we evaluate

$$(10.4)\qquad \left.\frac{\partial u}{\partial x}\right|_{x=x_j}=\sum_{n=0}^{N} a_n^{(1)}\cos\frac{\pi nj}{N}$$

using the fast Fourier transform. Finally, we evaluate $\exp(u_j+x_j)(\partial u/\partial x)_j$ at each of the "grid" points x_j and use the results to march the solution forward to the next time step.

For quadratically nonlinear differential equations, like the Navier–Stokes equations of incompressible fluid dynamics, Galerkin and tau approximations are workable but normally require at least twice the computational work of collocation approximation. However, Galerkin approximation is sometimes very attractive because it gives approximations that are conservative and have no so-called aliasing errors (see Orszag (1971c), (1972) for a more complete discussion of these properties). Energy conservation properties of spectral methods are discussed at the end of § 14.

Roundoff errors. Transform methods normally give no appreciable amplification of roundoff errors. In fact, the evaluation of convolution-like sums using fast Fourier transforms often gives results with much smaller roundoff error than would be obtained if the convolution sums were evaluated directly.

On the other hand, the use of recurrence relations to evaluate derivatives can sometimes be a source of large roundoff errors. In this case, it is often best to convert the problem being solved into a new one that is numerically well-conditioned. An example of such a transformation is given below.

EXAMPLE 10.1. *Solution of $y''-ky=f(x)$ by Chebyshev polynomials.* The boundary-value problem

$$(10.5)\qquad \begin{aligned} &y''-ky=f(x), \qquad -1\leqq x\leqq 1,\\ &y(-1)=A,\ y(1)=B \end{aligned}$$

can be solved using a Chebyshev–tau approximation. The resulting approximation $y_N(x)$ is given by (see Appendix)

$$(10.6)\qquad y_N(x)=\sum_{n=0}^{N} a_n T_n(x),$$

$$(10.7)\qquad \frac{1}{c_n}\sum_{\substack{p=n+2\\ p+n \text{ even}}}^{N} p(p^2-n^2)a_p-ka_n=f_n, \qquad 0\leqq n\leqq N-2,$$

$$(10.8)\qquad \sum_{n=0}^{N}(-1)^n a_n=A, \qquad \sum_{n=0}^{N} a_n=B,$$

where $\{f_n\}$ are the Chebyshev series coefficients of $f(x)$.

The solution of the system (10.7)–(10.8) for the Chebyshev coefficients $\{a_n\}$ may be done in several ways. One obvious way to do this efficiently is to write

$$(10.9)\qquad a_n=a_n^{(1)}+\alpha a_n^{(2)}+\beta a_n^{(3)}.$$

Here $a_n^{(1)}$ satisfies $a_N^{(1)}=a_{N-1}^{(1)}=0$ and

$$\frac{1}{c_n}\sum_{p=n+2}^{N} p(p^2-n^2)a_p^{(1)}-ka_n^{(1)}=f_n, \qquad 0\leqq n\leqq N-2,$$

while $a_n^{(2)}$ satisfies $a_N^{(2)}=1$, $a_{N-1}^{(2)}=0$ and

$$\frac{1}{c_n}\sum_{p=n+2}^{N} p(p^2-n^2)a_p^{(2)}-ka_n^{(2)}=0, \qquad 0\leqq n\leqq N-2,$$

and $a_n^{(3)}$ satisfies $a_N^{(3)}=0$, $a_{N-1}^{(3)}=1$, and

$$\frac{1}{c_n}\sum_{\substack{p=n+2\\ p+n \text{ even}}}^{N} p(p^2-n^2)a_p^{(3)}-ka_n^{(3)}=0, \qquad 0\leqq n\leqq N-2.$$

Each of the solutions $a_n^{(1)}$, $a_n^{(2)}$, $a_n^{(3)}$, may be found using roughly N operations by backwards recurrence. When the constants α and β in (10.9) are chosen so that the boundary conditions (10.8) are satisfied, a_n given by (10.9) satisfies (10.7)–(10.8).

The above procedure is efficient but it is not usually numerically stable. Roundoff errors multiply rapidly so that a_n may have little significance.

A better procedure is first to convert (10.7)–(10.8) into a nearly tridiagonal system of equations. It may be shown that (10.7)–(10.8) is *equivalent* to the system

$$\begin{aligned}(10.10)\qquad &\frac{kc_{n-2}}{4n(n-1)}a_{n-2}-\left(1+\frac{ke_{n+2}}{2(n^2-1)}\right)a_n+\frac{ke_{n+4}}{4n(n+1)}a_{n+2}\\ &=\frac{c_{n-2}f_{n-2}}{4n(n-1)}-\frac{e_{n+2}f_n}{2(n^2-1)}+\frac{e_{n+4}f_{n+2}}{4n(n+1)}, \qquad 2\leqq n\leqq N,\end{aligned}$$

with the boundary conditions (10.8) still applied. Here $c_0=2$, $c_n=1$ for $n>0$ and $e_n=1$ for $n\leqq N$, $e_n=0$ for $n>N$. The system (10.8), (10.10) may be solved by standard banded matrix techniques in roughly the number of operations required to solve pentadiagonal systems of equations. The equations in the form (10.10) are essentially diagonally dominant so no appreciable accumulation of roundoff errors occurs. This technique for solution of (10.5) is very useful in implementing implicit spectral methods for dissipative terms and for solving Poisson-like equations (see § 14).

SECTION 11

Numerical Results for Hyperbolic Problems

We begin by presenting numerical results for spectral approximations to the problem

$$u_t + u_x = 0, \qquad -1 \leqq x \leqq 1, \quad t > 0, \tag{11.1}$$

$$u(x, 0) = 0, \qquad u(-1, t) = g(t), \tag{11.2}$$

whose exact solution is

$$u(x, t) = \begin{cases} g(t-x-1), & x \leqq t-1, \\ 0, & x > t-1. \end{cases} \tag{11.3}$$

If $g(t)$ is smooth, $u(x, t)$ is smooth for $|x| < 1$ when $t > 2$; when $t < 2$, $u(x, t)$ is not smooth at $x = t-1$.

In § 2 we explained how to obtain semi-discrete Galerkin, tau, and collocation approximation to (11.1)–(11.2) using either Chebyshev or Legendre polynomial expansions. In § 9, we showed that either Adams–Bashforth or modified Euler time differencing gives stable and convergent results for these spectral approximations. The numerical results cited in this section were obtained by Adams–Bashforth time-differencing; time steps were chosen small enough that time-differencing errors are negligible.

Comparison of Chebyshev and Legendre polynomial spectral methods for smooth solutions. When $g(t) = -\sin M\pi t$, the solution (11.3) has M complete waves within $|x| \leqq 1$ when $t > 2$. As argued in § 3, we expect that accurate results will be obtained only if $N > M\pi$ polynomials are retained.

In Fig. 11.1, we plot the root-mean-square error for $|x| \leqq 1$ averaged in time for $4 \leqq t \leqq 4.4$ obtained using the Chebyshev approximations to (11.1)–(11.2) when $g(t) = -\sin 5\pi t$. In this time interval, $u(x, t)$ is smooth for $|x| \leqq 1$. Observe that the errors decrease exponentially fast when $N \gtrsim 5\pi$. Also observe that when the spectral approximations are relatively inaccurate (errors greater than roughly 10%), Galerkin approximation is most accurate followed by collocation and then tau. On the other hand, when the spectral approximations are very accurate (errors less than roughly 0.5%), tau approximation is most accurate followed by Galerkin and collocation. This behavior seems typical. Also observe from Fig. 11.1 that all three spectral approximations are nearly as accurate as the best (rms) Chebyshev approximation; in fact, tau approximation with $N+1$ polynomials is usually more accurate than the best approximation with N polynomials. Here the best (rms) Chebyshev approximation is that Nth degree polynomial that minimizes $\int_{-1}^{1} |u_N - u|^2 (1-x^2)^{-1/2}\, dx$.

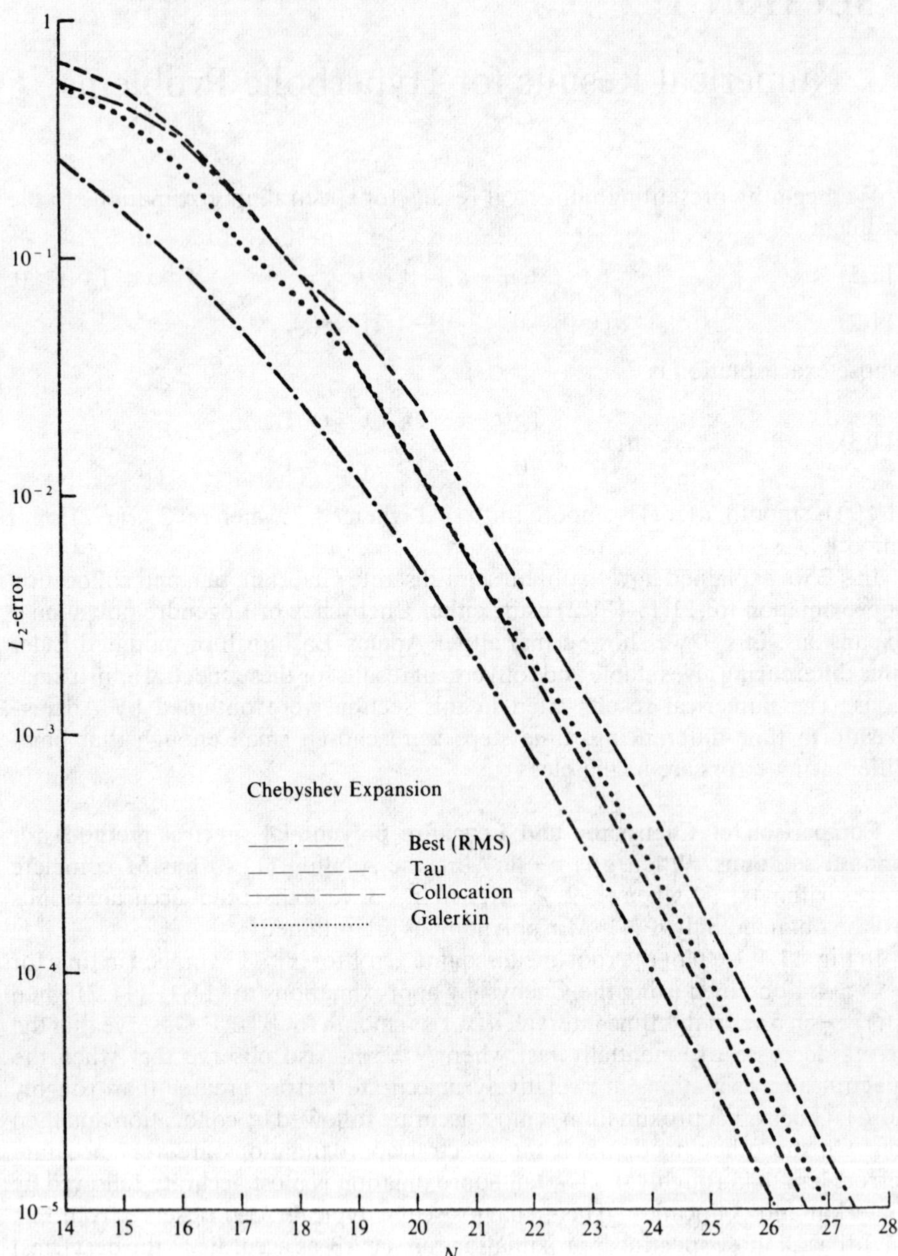

FIG. 11.1. *A plot of the L_2-errors in Chebyshev–spectral solution of* (11.1)–(11.2) *with $g(t) = -\sin 5\pi t$. The errors are averaged in time over the interval $4 \leqq t \leqq 4.4$; the exact solution $u(x, t) = \sin 5\pi(x + 1 - t)$ is smooth throughout this time interval. The best (rms) approximation is given by* (3.41) *with $M = 5$, $a = 1 - t$ truncated after $T_N(x)$. Observe that the errors decrease rapidly for $N > 5\pi$.*

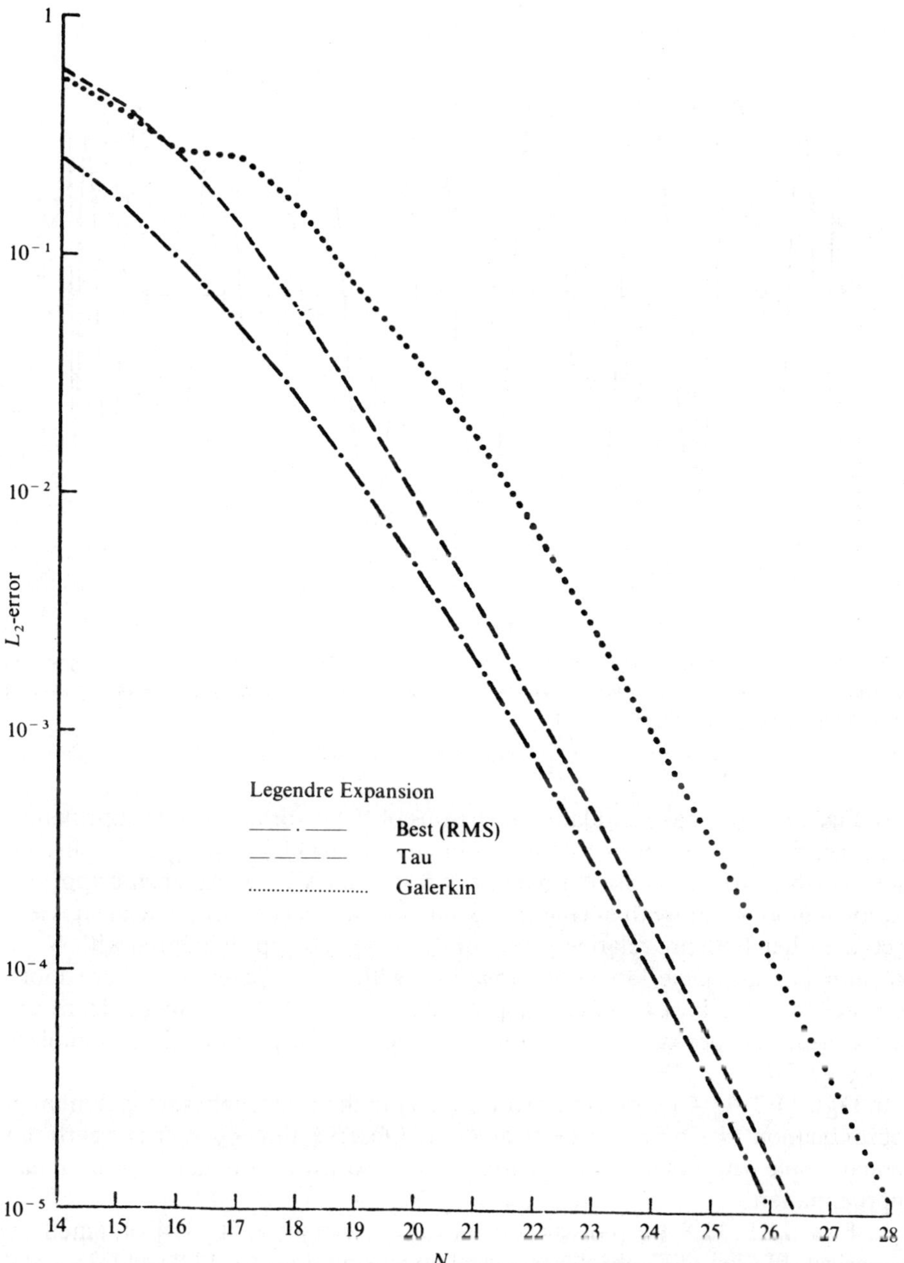

FIG. 11.2. *Same as Fig.* 11.1 *except for Legendre–spectral solution of* (11.1)–(11.2) *with* $g(t) = -\sin 5\pi t$. *Here the best* (*rms*) *approximation is given by* (3.45) *with* $M=5$, $a=1-t$ *truncated after* $P_N(x)$.

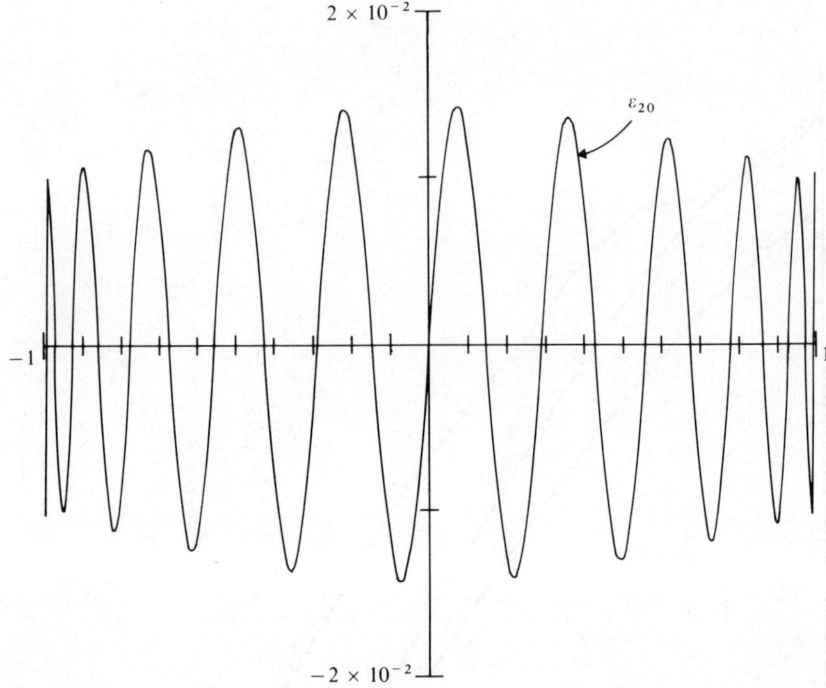

FIG. 11.3. *A plot of the error* $\varepsilon_N(X, t) = u_N(x, t) - u(x, t)$ *in the best (rms) Chebyshev polynomial approximation to* $u(x, t) = \sin 5\pi(x + 1 - t)$ *at* $t = 4$. *Here* $u_N(x, t) = \sum_{n=0}^{N} a_n(t)T_n(x)$ *with* $N = 20$ *and* $a_n(t) = (2/(\pi c_n)) \int_{-1}^{1} u(x, t)T_n(x)(1 - x^2)^{-1/2}\, dx$.

In Fig. 11.2, we make similar comparisons of the error in spectral approximations using Legendre series for the problem (11.1)–(11.2) with $g(t) = -\sin 5\pi t$. Here too the errors decrease exponentially fast when $N \gtrsim 5\pi$. Again, tau approximation is more accurate than Galerkin when both are very accurate, while it is less accurate when both are relatively inaccurate. Also, tau approximation with $N+1$ polynomials and Galerkin approximations with $N+2$ polynomials are more accurate than the best Legendre approximation with N polynomials. Here the best Legendre approximation is that Nth degree polynomial that minimizes $\int_{-1}^{1} |u_N - u|^2\, dx$.

In Figs. 11.3–11.4 we plot the error $\varepsilon_N(x, t)$ in the best Chebyshev polynomial approximation to $\sin 5\pi(x + 1 - t)$ at $t = 4$. Observe that $\varepsilon_N(x, t)$ is nearly an "equal ripple" approximation (Acton (1970)) so $u_N(x, t)$ is nearly a minimax approximation.

In Figs. 11.5–11.8 we plot the errors $\varepsilon_N(x, t)$ versus x at $t = 4$ obtained by numerical solution of Chebyshev spectral approximations to (11.1)–(11.2). As N increases, the tau method gives the closest approximation to an equal-ripple error, which is consistent with the result shown in Fig. 11.1 that tau approximation gives the smallest errors at high accuracy.

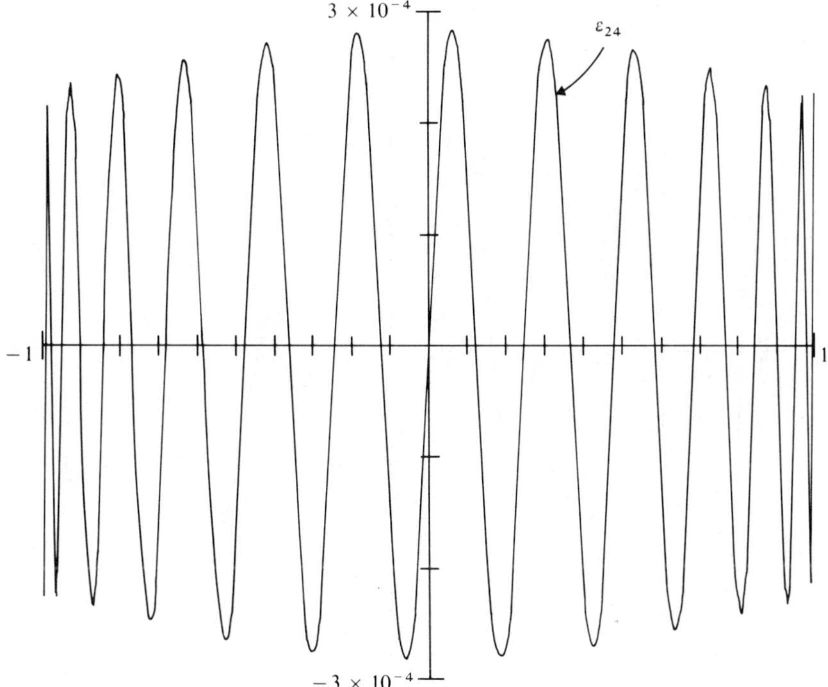

FIG. 11.4. *Same as Fig.* 11.3 *except* $N = 24$.

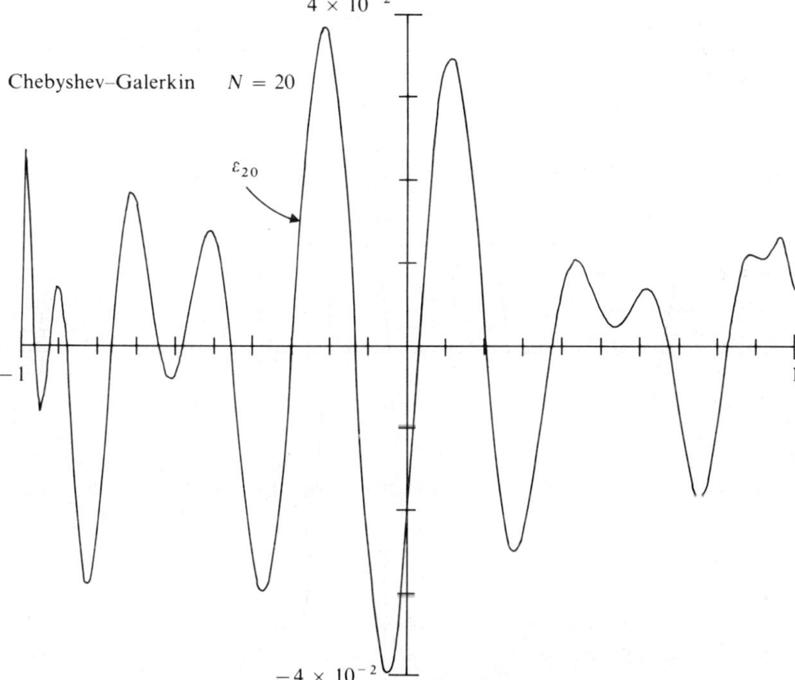

FIG. 11.5. *A plot of the error* $\varepsilon_N(x, t) = u_N(x, t) - u(x, t)$ *in the Chebyshev–Galerkin solution of* (11.1)–(11.2) *at* $t = 4$ *with* $g(t) = -\sin 5\pi t$. *Here* $u_N(x, t) = \sum_{n=0}^{N} a_n(t) T_n(x)$ *with* $N = 20$.

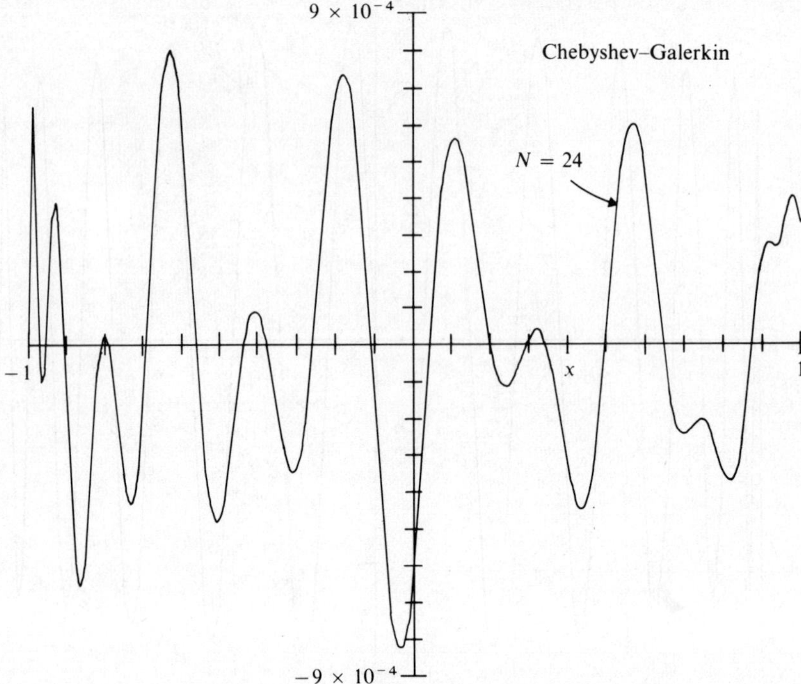

FIG. 11.6. *Same as Fig.* 11.5 *except* $N = 24$.

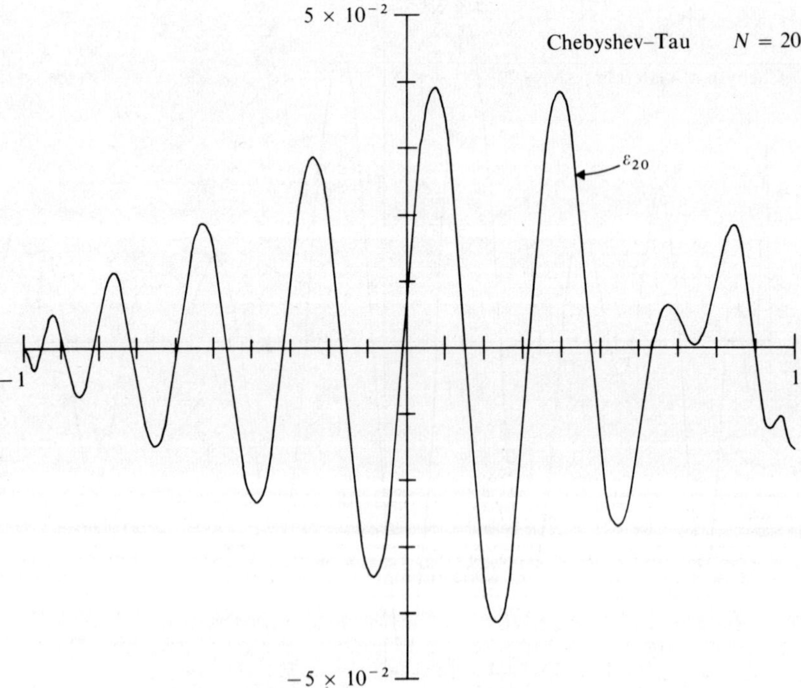

FIG. 11.7. *Same as Fig.* 11.5 *except that the Chebyshev–tau approximation to* (11.1)–(11.2) *is used with* $N = 20$.

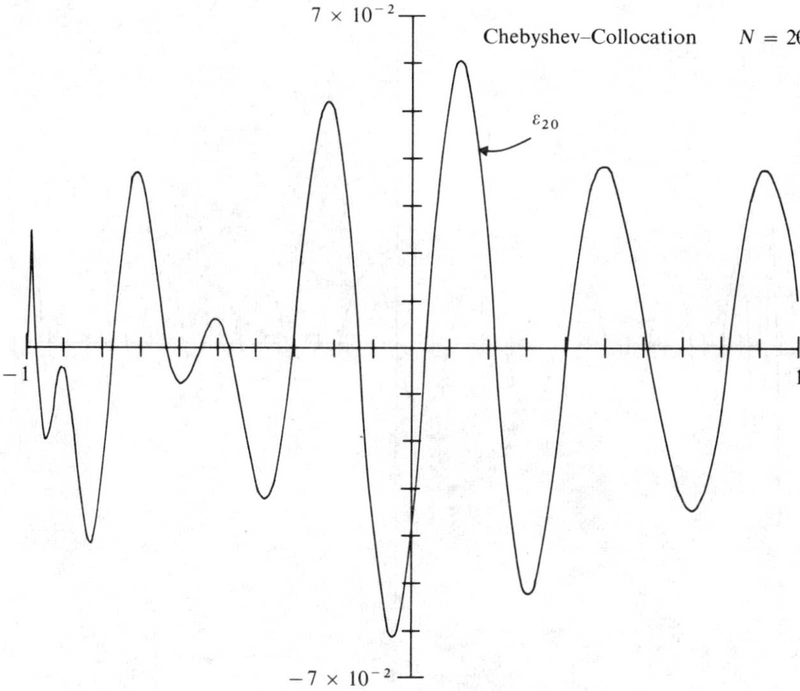

FIG. 11.8. *Same as Fig.* 11.5 *except that the Chebyshev collocation approximation to* (11.1)–(11.2) *is used with* $N = 20$.

In Figs. 11.9–11.10, we plot the error in the best Legendre polynomial approximation to $\sin 5\pi(x+1-t)$ at $t=4$. Observe that $\varepsilon_N(x, t)$ has large errors near the boundaries $x = \pm 1$. By comparing the results plotted in Figs. 11.3–11.4 with those plotted in Figs. 11.9–11.10, we conclude that the best Chebyshev polynomial approximation is closer to an equal ripple approximation than is the best Legendre polynomial approximation. Even though the best Legendre polynomial approximation to $u(x, t)$ gives the smallest mean-square error to u, the best Chebyshev polynomial approximation usually gives a smaller value of the maximum pointwise (L_∞) error. The large errors of the best Legendre approximation are concentrated near the boundaries $x = \pm 1$, while the Chebyshev weight function $(1-x^2)^{-1/2}$ tends to distribute the errors in the best Chebyshev approximation uniformly throughout $-1 \leqq x \leqq 1$.

In Figs. 11.11–11.13, we plot the errors $\varepsilon_N(x, t)$ at $t = 4$ obtained by numerical solution of Legendre spectral approximations to (11.1)–(11.2). As for Chebyshev–spectral approximations, the error in Legendre–tau approximation is smaller than that in Legendre–Galerkin approximation.

One important feature of Legendre–spectral approximation is that the spatial distribution of the error in tau and Galerkin approximation plotted in Figs. 11.11–11.13 differs markedly from the spatial distribution of the error in the best Legendre polynomial approximations plotted in Figs. 11.9–11.10. The boundary

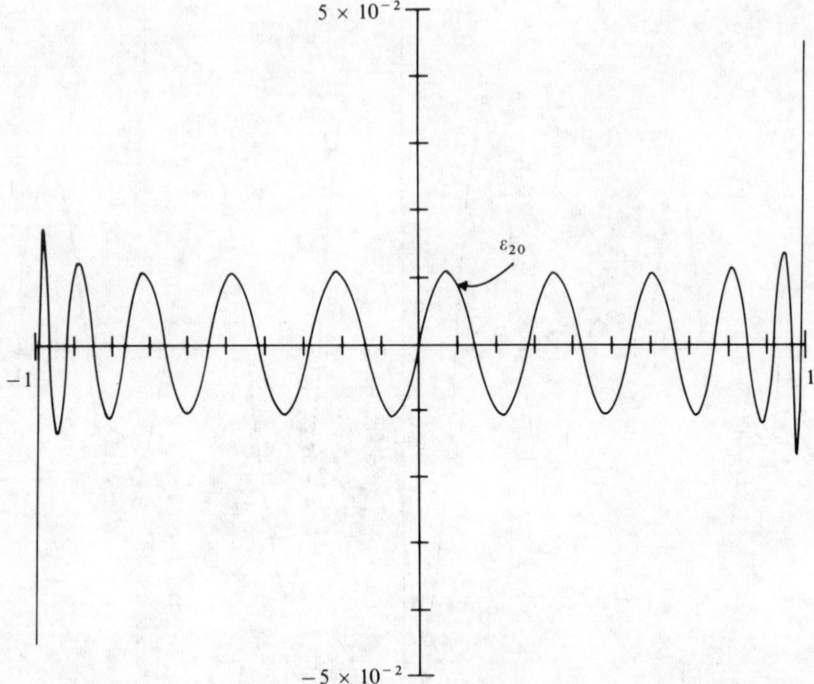

FIG. 11.9. *A plot of the error* $\varepsilon_N(x,t)=u_N(x,t)-u(x,t)$ *in the best (rms) Legendre polynomial approximation to* $u(x,t)=\sin 5\pi(x+1-t)$ *at* $t=4$. *Here* $u_N(x,t)=\sum_{n=0}^{N} a_n(t)P_n(x)$, $a_n=(n+\frac{1}{2})\int_{-1}^{1} u(x,t)P_n(x)\,dx$, *and* $N=20$. *Observe the large boundary errors of the Legendre polynomial approximation plotted here in comparison with the best Chebyshev polynomial approximation plotted in Fig.* 11.3.

errors in the best L_2 approximation are relatively large while the boundary errors are relatively smaller in the spectral approximations.

The boundary (endpoint) errors in Legendre–tau approximation exhibit "superconvergence" in the sense that they go to zero much faster than either the Legendre L_2-errors or the L_2 and endpoint errors of Chebyshev–tau approximation. This fact is illustrated in Fig. 11.14 where we plot the L_2 and endpoint errors of Legendre–tau and Chebyshev–tau spectral approximations to the solution of (11.1)–(11.2) with $g(t)=-\sin 5\pi t$. Here the endpoint error is $|u_N(+1,t)-u(+1,t)|$ at the outflow boundary $x=+1$.

Several features of the results plotted in Fig. 11.14 deserve comment. First, although the maximum error of the best N-term Chebyshev polynomial approximation is smaller than the maximum error of the best Legendre polynomial approximation to $u(x,t)$ by roughly a factor $1/\sqrt{N}$ [see (3.41) and (3.45)], the maximum error of the Legendre–tau approximation is smaller than the maximum error of the Chebyshev–tau approximation. Second, the endpoint error at $x=1$ of the Legendre–tau approximation goes to zero like the *square* of the endpoint error of the Chebyshev–tau approximation. This remarkable behavior of

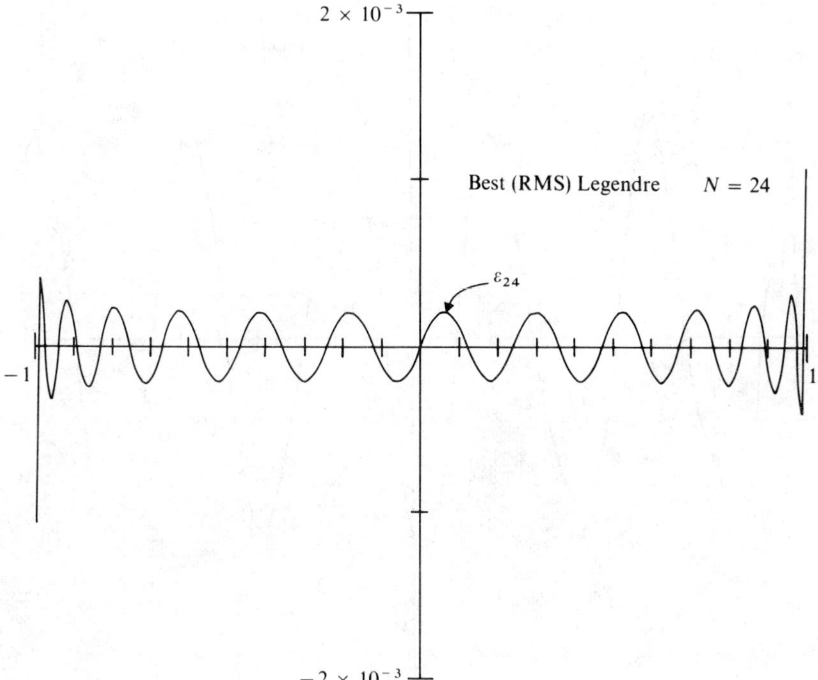

FIG. 11.10. *Same as Fig. 11.9 except* $N = 24$.

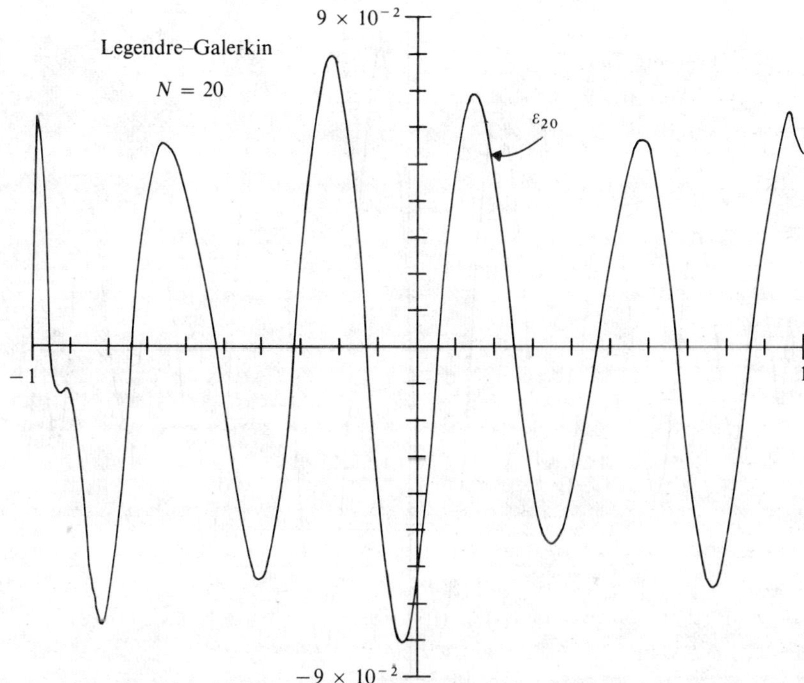

FIG. 11.11. *A plot of the error* $\varepsilon_N(x, t) = u_N(x, t) - u(x, t)$ *in the Legendre–Galerkin solution of* (11.1)–(11.2) *at* $t = 4$ *with* $g(t) = -\sin 5\pi t$. *Here* $u_N(x, t) = \sum_{n=0}^{N} a_n(t)P_n(x)$ *with* $N = 20$.

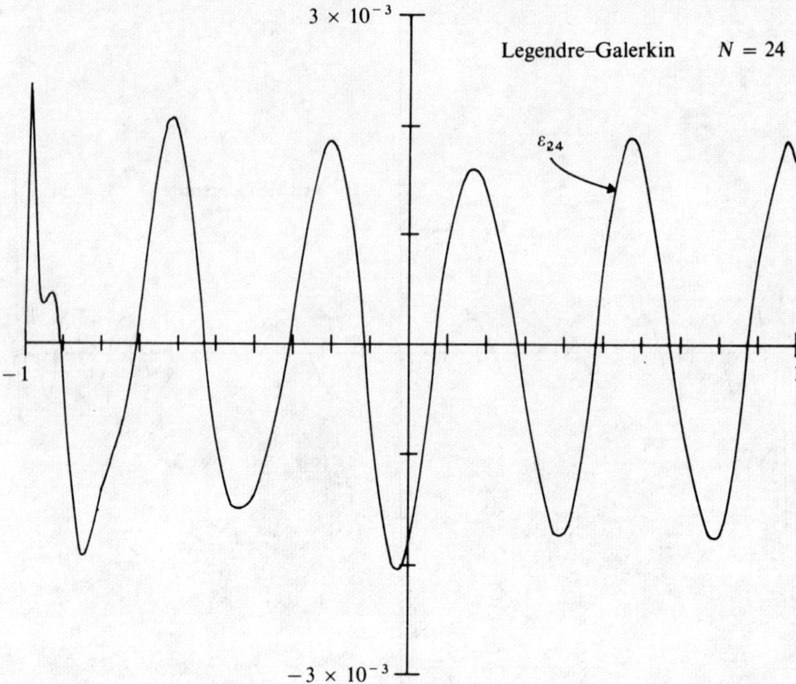

FIG. 11.12. *Same as Fig.* 11.11 *except* $N = 24$.

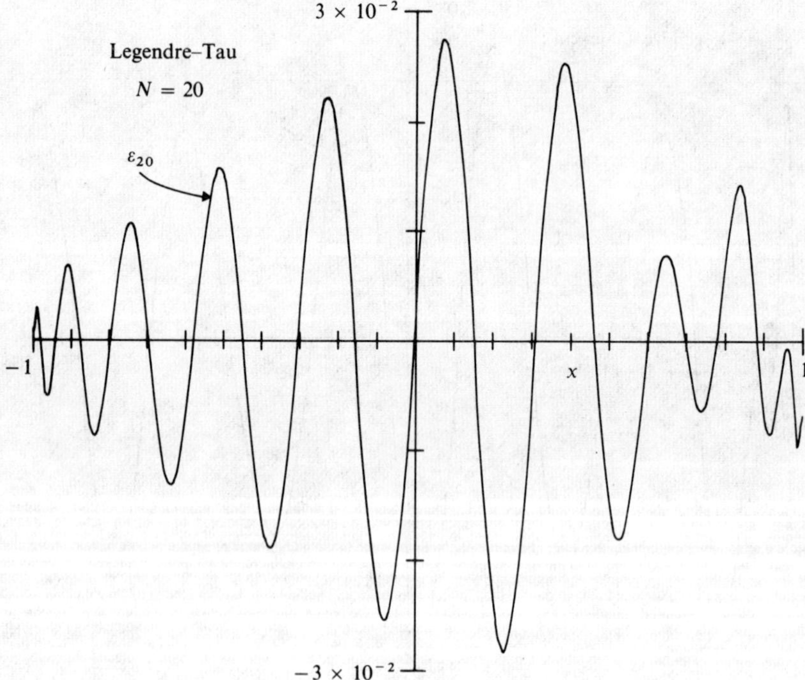

FIG. 11.13. *Same as Fig.* 11.11 *except that the Legendre–tau approximation to* (11.1)–(11.2) *is used with* $N = 20$.

endpoint errors in Legendre–polynomial approximations was found originally by Lanczos in a slightly different context (Lanczos (1966, p. 156), (1973)).

A mathematical analysis of the errors of spectral approximations to (11.1)–11.2) has been given recently by Dubiner (1977). Dubiner's results include: (a) asymptotic estimates of the errors incurred by the various spectral methods,

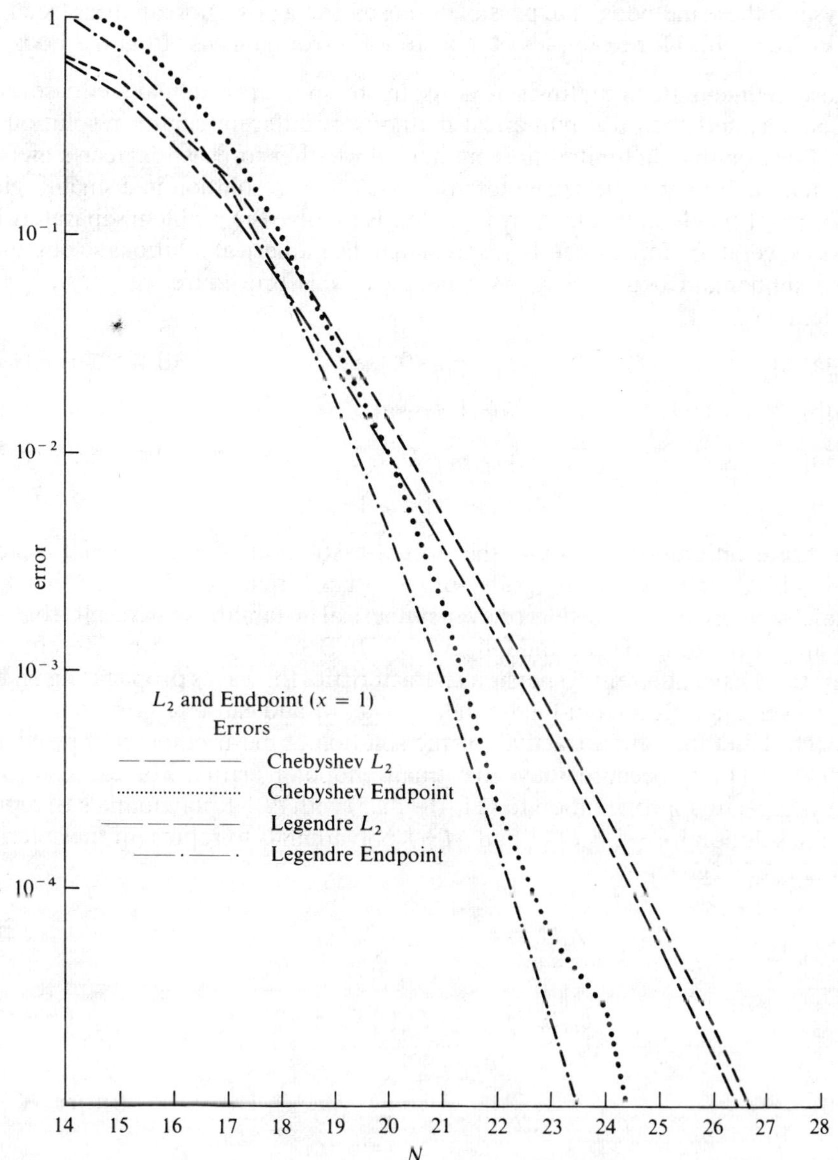

FIG. 11.14. *A comparison of the Chebyshev–tau and Legendre–tau L_2 and endpoint* ($x = +1$) *errors for the solution to* (11.1)–(11.2) *with* $g(t) = -\sin 5\pi t$.

including error oscillations when the solution is smooth; (b) a complete boundary layer description of the decay of large errors due to discontinuities after the discontinuities propagate out of the computational domain; (c) analysis of the behavior of the tau–function $\tau(t)$ in (2.35). Dubiner has analyzed a variety of spectral methods for (11.1)–(11.2) based on expansions in general Jacobi polynomials. His ingenious analyses of tau methods should permit more complete analysis of these methods than possible using earlier a posteriori analysis (see Fox and Parker (1968) for examples of a posteriori error analysis of tau methods).

Mesh refinement. Sometimes it is useful to split up a domain into several subdomains and then use numerical methods of different spatial resolution in each. For example, in limited-area numerical weather forecasting near a metropolitan area, it may be desirable to have much finer resolution in a small region than is practical globally. One way to do this is to solve the problem separately on each of several subdomains and then to match the numerical solutions so obtained across subdomain boundaries. As a model of this procedure we consider the problem

$$u_t + u_x = 0, \qquad -1 \leqq x \leqq 1, \quad t > 0, \tag{11.4a}$$

$$u(-1, t) = g(t), \tag{11.4b}$$

$$v_t + v_x = 0, \qquad 1 \leqq x \leqq 3, \quad t > 0, \tag{11.5a}$$

$$v(1+, t) = u(1-, t). \tag{11.5b}$$

With finite difference methods, the accurate solution of the coupled system (11.4)–(11.5) using different grids for $-1 \leqq x \leqq 1$ than for $1 \leqq x \leqq 3$ may be troublesome. Inaccurate results or even numerical instabilities can result from the matching (Browning, Kreiss and Oliger (1973)). Because grids with different grid separations have different dispersion characteristics for waves propagating on the grid, waves can reflect from the boundary at $x = 1$ and cause large errors.

Spectral methods are attractive for the solution of mesh refinement problems like (11.4)–(11.5) because they give small endpoint errors. For example, the Chebyshev–tau approximation to (11.4)–(11.5) with $N+1$ polynomials to represent the solution for $-1 \leqq x \leqq 1$ and $M+1$ polynomials to represent the solution for $1 \leqq x \leqq 3$ is given by

$$u_N(x, t) = \sum_{n=0}^{N} a_n(t) T_n(x), \qquad -1 \leqq x \leqq 1, \tag{11.6}$$

$$v_M(x, t) = \sum_{m=0}^{M} b_m(t) T_m(x-2), \qquad 1 \leqq x \leqq 3, \tag{11.7}$$

$$\frac{da_n}{dt} = -\frac{2}{c_n} \sum_{\substack{p=n+1 \\ p+n \text{ odd}}}^{N} p a_p, \qquad 0 \leqq n \leqq N-1, \tag{11.8}$$

$$\frac{db_m}{dt} = -\frac{2}{c_m} \sum_{\substack{p=m+1 \\ p+m \text{ odd}}}^{M} p b_p, \qquad 0 \leqq m \leqq M-1, \tag{11.9}$$

$$\sum_{n=0}^{N} (-1)^n a_n = g(t), \tag{11.10}$$

$$\sum_{m=0}^{M} (-1)^m b_m = \sum_{n=0}^{N} a_n. \tag{11.11}$$

It may easily be shown that if $g(t)$ is smooth, the solution to (11.6)–(11.11) converges to the solution of (11.4)–(11.5) throughout $-1 \leqq x \leqq 3$ faster than any finite power $1/N$ or $1/M$ as $N, M \to \infty$. The solutions for $-1 \leqq x \leqq 1$ and $1 \leqq x \leqq 3$ match without the necessity of imposing any matching conditions in addition to (11.5b) or (11.11). Because no spurious downstream boundary conditions are applied at $x = +1$ on the wave propagating in the interval $-1 \leqq x \leqq 1$, there are no reflected waves.

One more example of a refined mesh spectral calculation is instructive. Consider the heat equation problem

$$\frac{\partial u}{\partial t} = \nu \frac{\partial^2 u}{\partial x^2}, \qquad -1 \leqq x \leqq 1, \tag{11.12a}$$

$$\frac{\partial v}{\partial t} = \nu \frac{\partial^2 v}{\partial x^2}, \qquad 1 \leqq x \leqq 3, \tag{11.12b}$$

$$u(-1, t) = f(t), \qquad v(3, t) = g(t) \tag{11.12c}$$

$$u(1-, t) = v(1+, t), \qquad \frac{\partial u}{\partial x}(1-, t) = \frac{\partial v}{\partial x}(1+, t) \tag{11.12d}$$

where (11.12d) follows by requiring continuity of temperature and heat flux across the boundary at $x = 1$. A Chebyshev–tau approximation to (11.12) is given by (11.6)–(11.7) with

$$\frac{da_n}{dt} = \frac{\nu}{c_n} \sum_{\substack{p=n+2 \\ p+n \text{ even}}}^{N} p(p^2 - n^2) a_p, \qquad 0 \leqq n \leqq N-2, \tag{11.13a}$$

$$\frac{db_m}{dt} = \frac{\nu}{c_m} \sum_{\substack{p=m+2 \\ p+m \text{ odd}}}^{M} p(p^2 - m^2) b_p, \qquad 0 \leqq p \leqq M-2, \tag{11.13b}$$

$$\sum_{n=0}^{N} (-1)^n a_n = f(t), \qquad \sum_{m=0}^{M} b_m = g(t), \tag{11.13c}$$

$$\sum_{n=0}^{N} a_n = \sum_{m=0}^{M} (-1)^m b_m, \qquad \sum_{n=0}^{N} n^2 a_n = -\sum_{m=0}^{M} (-1)^m m^2 b_m. \tag{11.13d}$$

It may be shown as in Example 7.1(v) that this approximation is semi-bounded and hence stable and convergent.

Discontinuities. When $t < 2$, the solution (11.3) to (11.1)–(11.2) is not smooth at $x = t - 1$; if $g(t) = \sin M\pi t$, the solution has a discontinuous derivative. This discontinuity seriously degrades the rate of convergence of spectral approximations near the discontinuity. Nevertheless, spectral approximations are still normally much more accurate than finite-difference approximations to the same problem. Orszag and Jayne (1974) give comparisons between finite-difference and spectral approximations to discontinuous solutions; in particular, they argue that if the pth derivative of the solution is discontinuous, the rate of convergence of Chebyshev–spectral approximations to (11.1)–(11.3) for $t<2$ is of order $1/N^p$ as $N\to\infty$. Dubiner (1977) has given a detailed asymptotic analysis of this problem. His results include detailed behavior of the error for all x and t and are in good agreement with numerical solutions.

One of the attractive features of spectral methods for problems with discontinuities is that the region of large errors is more localized near the discontinuity than in finite-difference methods. Thus, it should be possible to eliminate oscillations near the discontinuity using less dissipation than is required when finite difference methods are used. A comparison of the error in Chebyshev–tau and second- and fourth-order finite-difference solutions of (11.1)–(11.2) for $t<2$ is given in Fig. 11.15.

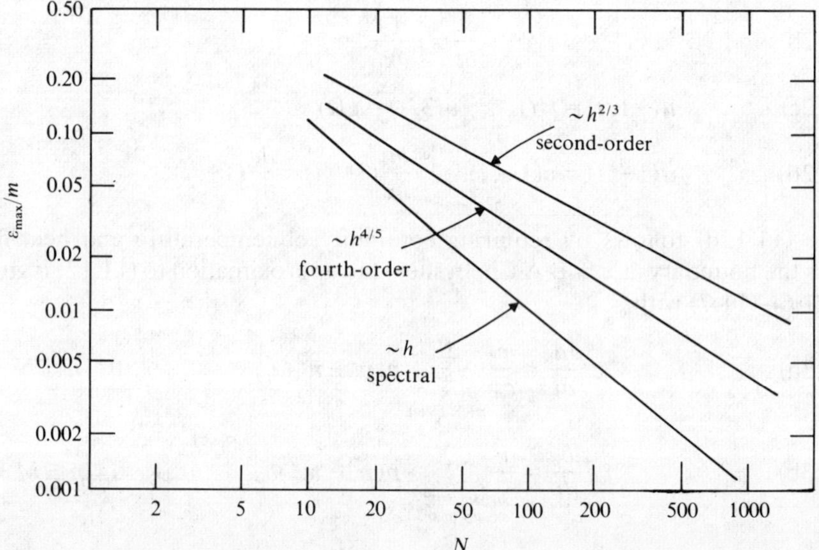

FIG. 11.15. *Maximum pointwise errors in the solution of* (11.1)–(11.2) *with* $g(t)=\sin m\pi t$ *at* $t=1.2$ *when the discontinuity in the solution* (11.3) *is at* $x=0.2$. *Here* ε_{max} *is the maximum error and* $N=2/h$ *is the number of grid points, or of Chebyshev polynomials in the spectral method.*

Another interesting way to use spectral methods for problems with discontinuous solutions has been suggested by Boris and Book (1976). The "optimal flux-corrected transport" approximation gives good resolution of discontinuities without introduction of unphysical numerical oscillations near the discontinuity. The idea is to add in an artificial diffusion to smooth the discontinuity and then to "anti-diffuse" the resulting solution in such a way that no new oscillations or maxima/minima are produced.

Comparison with finite-difference methods. Finite-difference approximations to (11.1)–(11.2) must be formulated carefully near the boundaries $x = \pm 1$. For example, the fourth-order semi-discrete approximation

$$\frac{\partial u_j}{\partial t}+\frac{8(u_{j+1}-u_{j-1})-u_{j+2}+u_{j-2}}{12\,\Delta x}=0$$

where $u_j(t)=u(j\,\Delta x, t)$, must be modified at $x=-1+\Delta x,\ 1-\Delta x, 1$ because $u(-1-\Delta x, t), u(1+\Delta x, t), u(1+2\,\Delta x, t)$ all lie outside the computational domain $-1 \leqq x \leqq 1$. Kreiss and Oliger (1973) discuss methods to formulate difference approximations at these grid points. However, it is not known how to formulate appropriate "boundary" conditions for arbitrary order difference schemes. This difficulty is an artifact of difference methods; a fourth-order difference equation requires 4 "boundary" conditions while only 1 condition (11.2) is properly imposed on the first-order differential equation (11.1).

On the other hand, properly formulated spectral methods require no "spurious" boundary conditions. Indeed, the imposition of a spurious boundary condition on a spectral approximation to (11.1), like $\partial u/\partial x = 0$ at $x = +1$, will induce an unconditional instability (see §§ 8 and 12). The mathematics of spectral approximations follows closely the mathematics of the differential equation being solved.

Spectral approximations also require considerably fewer degrees of freedom to achieve accurate results than are required by difference methods. A comparison for the problem (11.1)–(11.2) is given in Table 11.1 for late times at which the solution is smooth.

TABLE 11.1

L_2 (rms) errors for the solution of (11.1)–(11.2) with $g(t)=\sin M\pi t$. The errors listed are measured at $t=5$ when the solution (11.3) is smooth. Time differencing errors are negligible and N is the number of grid points or Chebyshev polynomials. Observe that to achieve a 1% error, the second-order method requires $N/M \gtrsim 40$, the fourth-order method requires $N/M \gtrsim 15$, while the Chebyshev–tau method requires $N/M \gtrsim \pi$.

Second-order			Fourth-order			Chebyshev–tau		
N	M	ε_2	N	M	ε_4	N	M	ε_∞
40	2	0.1	20	2	0.04	16	4	0.08
80	2	0.03	30	2	0.008	20	4	0.001
160	2	0.008	40	2	0.002	28	8	0.2
40	4	1	40	4	0.07	32	8	0.008
80	4	0.2	80	4	0.005	42	12	0.2
160	4	0.06	160	4	0.0003	46	12	0.02

In Figs. 11.16–19 we show three-dimensional perspective plots of the solution to a simple two dimensional hyperbolic problem with periodic boundary conditions:

$$\frac{\partial A(x,y,t)}{\partial t}-y\frac{\partial A(x,y,t)}{\partial x}+x\frac{\partial A(x,y,t)}{\partial y}=0 \tag{11.14}$$

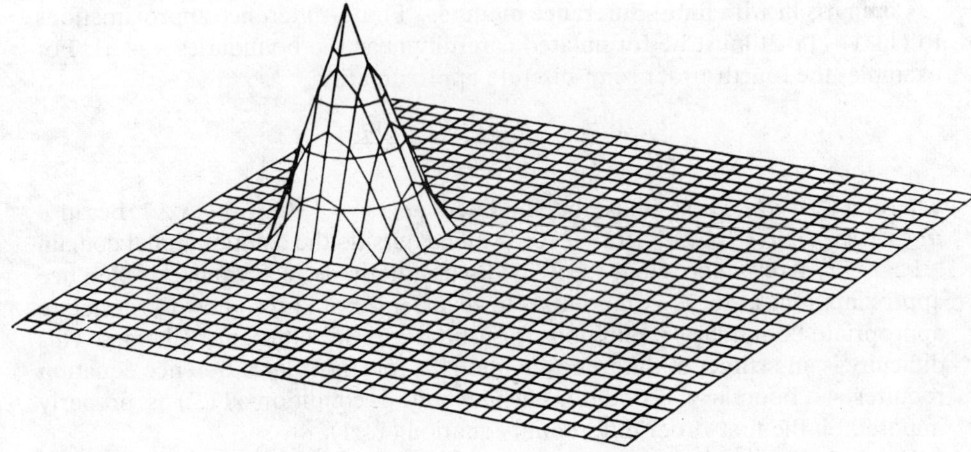

FIG. 11.16. *A perspective plot of the* $A(x, y, t=0)$ *used in a numerical test of methods to solve* (11.14).

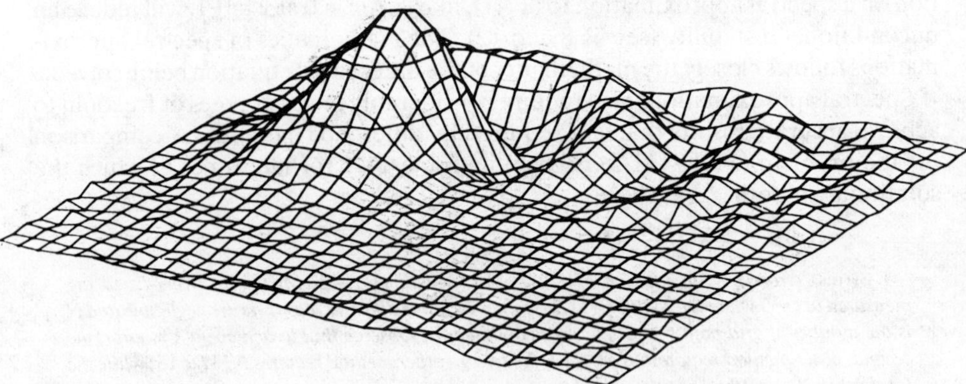

FIG. 11.17. *Three-dimensional perspective plot of the* $A(x, y, t)$ *field obtained after one revolution using a second-order centered difference scheme on a* 32×32 *grid.*

with

$$A(x\pm 2\pi, y\pm 2\pi, t)=A(x, y, t).$$

The solution to (11.14) is constant along the characteristics $x+iy=(x_0+iy_0)\,e^{it}$. Therefore, $A(x, y, 2\pi)\equiv A(x, y, 0)$ so the solution should keep A unchanged after a time 2π. In Fig. 11.16, we plot the initial conditions used for the calculation whose results are plotted in Figs. 11.17–19. In Fig. 11.17 we plot the results at $t=2\pi$ of a second-order centered space difference scheme; in Fig. 11.18 we plot the results of a fourth-order scheme and in Fig. 11.19 we plot the results of a spectral calculation using the Fourier expansion

$$A(x, y, t)=\sum_{|k|\leqq K}\sum_{|p|\leqq P} a(k, p, t)\,e^{ikx+ipy}.$$

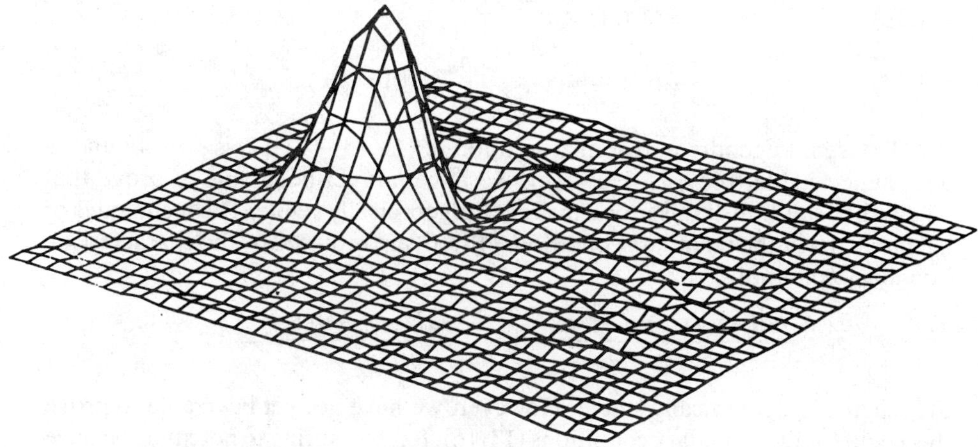

FIG. 11.18. *Same as Fig.* 11.17 *except using a fourth-order centered difference scheme on a* 32×32 *grid.*

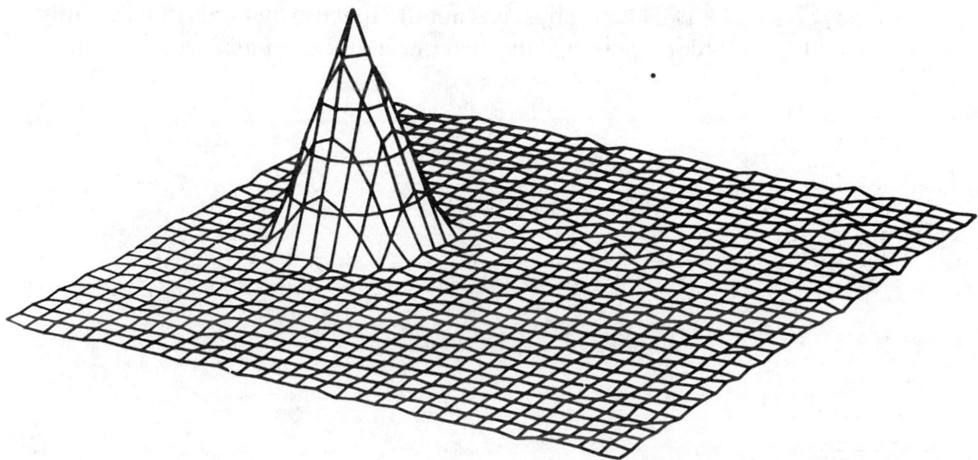

FIG. 11.19. *Same as Fig.* 11.17 *except using a Fourier–spectral method with* $K=P=16$ (32×32 *modes*).

All three calculations used the same number of degrees of freedom but the differences in accuracy are striking. The Fourier–spectral method works well even though the convecting velocity $(-y, x)$ in (11.14) has jump discontinuities at $x=\pm 2\pi$, $y=\pm 2\pi$. The dominant error in all three calculations originates from the "corners" of the cone in the initial $A(x, y, 0)$ distribution; this error appears as a large lagging phase error in the finite difference solutions which explains the "wakes" of large errors following the remnants of $A(x, y, 2\pi)$.

Higher-order wave equations. The mixed initial-boundary value problem

$$\frac{\partial^2 u}{\partial t^2}=\frac{\partial^2 u}{\partial x^2}, \qquad -1\leqq x\leqq 1, \quad t>0, \tag{11.15}$$

$$u(\pm 1, t) = 0, \tag{11.16}$$

$$u(x, 0) = f(x), \qquad \frac{\partial u}{\partial t}(x, 0) = g(x) \tag{11.17}$$

is well posed. Legendre polynomial solution of (11.15)–(11.17) is semi-bounded and, hence stable (see § 7). However, we have not yet been able to prove that Chebyshev solution of this problem is even algebraically stable. The techniques of § 8 prove that if the boundary conditions (11.16) are replaced by the characteristic boundary conditions

$$\frac{\partial u(-1, t)}{\partial t} - \frac{\partial u(-1, t)}{\partial x} = 0, \qquad \frac{\partial u(1, t)}{\partial t} + \frac{\partial u(1, t)}{\partial x} = 0,$$

the scheme is algebraically stable. However, we have not yet been able to prove this result for the boundary conditions (11.16). It is reassuring to note that we have solved the Chebyshev–spectral approximations to (11.15)–(11.17) and find no evidence of lack of convergence. Indeed, the Chebyshev methods work just as well as they do for (11.1)–(11.2). Thus, it is not the spectral methods that run into difficulty on higher-order equations, but just our methods of analysis.

SECTION 12

Advection-Diffusion Equation

In this section, we consider spectral methods for the advection-diffusion ("linearized Burgers") equation

$$\frac{\partial u(x,t)}{\partial t}+U\frac{\partial u(x,t)}{\partial x}=\nu\frac{\partial^2 u}{\partial x^2}+f(x,t), \qquad -1\leqq x\leqq 1, \tag{12.1}$$

$$u(-1,t)=0, \qquad u(1,t)=0, \tag{12.2}$$

$$u(x,0)=g(x). \tag{12.3}$$

Equation (12.1) is parabolic so boundary conditions should be applied at both $x=-1$ and $x=+1$. When ν is small, the boundary condition applied at $x=+1$ (assuming $U>0$) has an interesting effect on the stability of the spectral methods. To begin, we remark that the analyses of §§ 7–8 can be extended to show that, as $N\rightarrow\infty$, N-term Legendre and Chebyshev approximations to (12.1)–(12.3) are stable and convergent.

For example, Chebyshev–Galerkin approximation is stable because (12.1)–12.2) and (7.3) imply that

$$\begin{aligned}\frac{d}{dt}\int_{-1}^{1}\frac{u^2}{\sqrt{1-x^2}}\,dx &\leqq |U|\int_{-1}^{1}\frac{u^2}{(1-x^2)^{3/2}}\,dx-\nu\int_{-1}^{1}\frac{u^2}{(1-x^2)^{5/2}}\,dx\\ &\leqq |U|\int_{-\sqrt{1-\nu/U}}^{\sqrt{1-\nu/U}}\frac{u^2}{(1-x^2)^{3/2}}\,dx\\ &\leqq\frac{U^2}{\nu}\int_{-1}^{1}\frac{u^2}{\sqrt{1-x^2}}\,dx\end{aligned} \tag{12.4}$$

so the approximation is semi-bounded.

However, for finite N, there may be difficulty integrating the resulting spectral equations. With Legendre polynomials, Galerkin approximation L_N to (12.1)–(12.3) satisfies $L_N+L_N^*\leqq 0$ so there is no difficulty with time integrations (although the solution may not be accurate unless N is large enough.

On the other hand, Chebyshev–spectral solution of (12.1)–(12.3) encounters the following curious behavior when ν is small. If ν/U is small and N is not too large, the Chebyshev–spectral approximations L_N to (12.1)–(12.3) have eigenvalues with positive real parts. In Table 12.1, we list values of N_{crit} for various values of ν/U; for $N<N_{\text{crit}}$, L_N for Chebyshev–tau approximation to (12.1)–(12.3) has eigenvalues with positive real parts. Since these eigenvalues may have moderately large real parts (they can be as large as $U^2/(2\nu)$ by (12.4)], there may be rapid growth of errors and numerical solution of the Chebyshev–spectral

equations may appear unstable and divergent. For $N \geqq N_{\text{crit}}$, there are no eigenvalues of L_N with positive real parts so the spectral equations are stable in t.

The origin of this temporal instability is the outflow boundary layer at $x = \pm 1$; when $U > 0$, the solution to (12.1)–(12.3) develops a region of rapid change of width roughly ν/U near $x = +1$ as t increases. Since roughly $3(U/\nu)^{1/2}$ Chebyshev polynomials are required to resolve a boundary layer of width ν/U [see (3.50)], we expect that $N_{\text{crit}} \doteqdot 3(U/\nu)^{1/2}$ so $\nu N_{\text{crit}}^2/U \doteqdot 9$. In fact, as shown in Table 12.1, the criterion is actually $\nu N_{\text{crit}}^2/U \doteqdot 4$. (Since the Chebyshev norm of $\exp(-Ut\,\partial/\partial x)$ is roughly $N^{1/4}$ (see § 8), we expect that the proper scaling of N_{crit} is better represented as $\nu N_{\text{crit}}^{7/4}/U \doteqdot 1.3$. As shown in Table 12.1, this modified scaling is more nearly satisfied for the range of ν considered.)

TABLE 12.1

Critical values N_{crit} of the number of Chebyshev polynomials necessary that the tau approximation to the operator $-U\,\partial u/\partial x + \nu\,\partial^2 u/\partial x^2$ with $u(\pm 1) = 0$ have no eigenvalue with positive real parts. Also listed are the inverse "grid Reynolds number" $\nu N_{\text{crit}}^2/U$ and the parameter $\nu N_{\text{crit}}^{7/4}/U$.

ν/U	N_{crit}	$\nu N_{\text{crit}}^2/U$	$\nu N_{\text{crit}}^{7/4}/U$
1.0×10^{-2}	15	2.25	1.14
2.5×10^{-3}	35	3.06	1.26
1.0×10^{-3}	61	3.72	1.33
6.0×10^{-4}	81	3.94	1.31
4.0×10^{-4}	101	4.08	1.29

If Chebyshev–spectral approximations to (12.1)–(12.3) are solved using fractional time-step methods, the temporal instability for $N < N_{\text{crit}}$ appears in a unique way. Define the operator A_N as an N-mode Chebyshev approximation to the operator $-U\,\partial u/\partial x$ with the boundary condition $u(-1) = 0$ and the operator B_N as an N-mode Chebyshev approximation to the operator $\nu\,\partial^2 u/\partial x^2$ with $u(\pm 1) = 0$. Then the evolution operator of (12.1)–(12.2) is $\exp[(A_N + B_N)t]$ so a fractional step method involves the splitting $\partial u_N/\partial t = \partial_1 u_N/\partial t + \partial_2 u_N/\partial t$ where

$$\partial_1 u_N/\partial t = A_N u_N, \qquad \partial_2 u_N/\partial t = B_N u_N.$$

For any values of ν and $U > 0$, the fractional step $\partial_1 u_N/\partial t$ is algebraically stable since $\|\exp A_N t\| = O(N^{1/4})$ (see § 8), while the fractional step $\partial_2 u_N/\partial t$ is stable since $\|\exp B_N t\| \leqq 1$ (see § 7). Nevertheless, $\|\exp[(A_N + B_N)t]\|$ can grow rapidly with t. The reason is that A_N and B_N do not commute so it is not true that $\|\exp[(A_N + B_N)t]\| \leqq \|\exp A_N t\|\,\|\exp B_N t\|$. The Lie formula (5.8) does ensure that

$$\|\exp[(A_N + B_N)t]\| \leqq \lim_{n\to\infty} \|\exp(A_N t/n)\|^n \|\exp(B_N t/n)\|^n.$$

However, as $N \to \infty$, $n/N \to \infty$

$$\|\exp(A_N t/n)\| - 1 \sim cN^2 t/n$$

with $c>0$ (see Table 8.1) so

$$\|\exp(A_N t/n)\|^n \sim \exp(cN^2 t) \gg 1, \qquad N\to\infty, \quad n/N^2\to\infty.$$

Therefore the Lie formula gives only the very weak upper bound

$$\|\exp[(A_N+B_N)t]\| \leqq \exp(cN^2 t).$$

In summary, Chebyshev–spectral approximations to (12.1)–(12.3) give fractional step methods such that each fractional step is algebraically stable while the total step is unstable unless $N>N_{\text{crit}}$.

If the boundary conditions (12.2) are replaced by

$$u(-1,t)=0, \qquad \frac{\partial u}{\partial x}(+1,t)=0 \tag{12.5}$$

when $U>0$, the criterion for temporal stability is relaxed significantly. As shown in Table 12.2, the value of $\nu N_{\text{crit}}^2/U$ is decreased to roughly 1.6. However, the growing modes that appear when $N<N_{\text{crit}}$ are much tamer than those that appear when the boundary condition $u(+1,t)=0$ is applied, so accurate time integrations are still practicable when $\nu N^2/U \approx 0.01$ (see Haidvogel (1977)).

TABLE 12.2

Critical values N_{crit} of the number of Chebyshev polynomials necessary that the tau approximation to the operator $-U\,\partial u/\partial x+\nu\,\partial^2 u/\partial x^2$ with $u(-1)=0$, $\partial u(+1)/\partial x=0$ and $U>0$ have no eigenvalues with positive real parts. The parameter $\nu N_{\text{crit}}^{7/4}/U$ is also listed.

ν/U	N_{crit}	$\nu N_{\text{crit}}^{7/4}/U$
2.5×10^{-3}	21	0.52
1.0×10^{-3}	37	0.56
6.0×10^{-4}	49	0.54
4.0×10^{-4}	61	0.53
2.0×10^{-4}	89	0.52

SECTION 13

Models of Incompressible Fluid Dynamics

The Stokes equations for low Reynolds number, two-dimensional incompressible flow are

$$\frac{\partial \mathbf{v}}{\partial t} = -\nabla p + \nu \nabla^2 \mathbf{v}, \qquad \nabla \cdot \mathbf{v} = 0, \tag{13.1}$$

where $\mathbf{v}$ is the velocity field, p is the pressure, and ν is the kinematic viscosity. With the boundary conditions that $\mathbf{v}=0$ on rigid stationary boundaries, the problem (13.1) is well posed for any $\nu>0$. An equivalent formulation is given by the vorticity–streamfunction equations

$$\frac{\partial \zeta}{\partial t} = \nu \nabla^2 \zeta, \qquad \zeta = \nabla^2 \psi, \tag{13.2}$$

obtained by taking the curl of the Stokes equations (13.1). Here ψ is the streamfunction defined by $\mathbf{v}=(-\partial\psi/\partial y, \partial\psi/\partial x)$ and ζ is the vorticity.

A one-dimensional model of (13.2) is

$$\frac{\partial \zeta}{\partial t} = \nu \frac{\partial^2 \zeta}{\partial x^2}, \qquad -1 \leqq x \leqq 1, \quad t>0, \tag{13.3}$$

$$\zeta = \frac{\partial^2 \psi}{\partial x^2}, \qquad -1 \leqq x \leqq 1. \tag{13.4}$$

On stationary rigid walls, the boundary conditions for (13.3)–(13.4) are

$$\psi(x,t) = \psi_x(x,t) = 0, \qquad x = \pm 1. \tag{13.5}$$

There is one subtlety in the application of spectral methods to (13.3)–(13.5) that does not appear directly when the primitive equations (13.1) are used. It is necessary to use some care to avoid unconditional numerical instability with the Chebyshev tau method.

The most obvious way to use the tau method to solve (13.3)–(13.5) is to substitute (13.4) into (13.3) and solve

$$\psi_{xxt} = \nu \psi_{xxxx}, \qquad -1 \leqq x \leqq 1, \quad t>0, \tag{13.6}$$

by expanding $\psi(x,t)$ in the Chebyshev series

$$\psi(x,t) = \sum_{n=0}^{N} a_n(t) T_n(x). \tag{13.7}$$

Denoting by $a_n^{(q)}$ the Chebyshev expansion coefficients of $\partial^q\psi/\partial x^q$ (see (A.23)), the tau equations for (13.5)–(13.6) are

$$\frac{da_n^{(2)}}{dt}=\nu a_n^{(4)}, \qquad 0\leqq n\leqq N-4, \quad t>0, \tag{13.8}$$

$$\sum_{n=0}^{N}(\pm 1)^n a_n=\sum_{n=0}^{N}(\pm 1)^n n^2 a_n=0. \tag{13.9}$$

Unfortunately, this method for solution of (13.3)–(13.5) is unconditionally unstable as $N\to\infty$. In Table 13.1, we list the largest positive eigenvalue $\lambda_{\max}$ of (13.8)–(13.9); there is a solution of (13.8)–(13.9) that grows like $a_n(t)=a_n(0)\exp(\lambda_{\max}t)$. Since $\lambda_{\max}$ grows like N^4 as $N\to\infty$, errors also grow rapidly as $N\to\infty$ for fixed t. This method is unuseable for time-dependent calculations.

In Table 13.1, we also list the values of λ_n for $n=1, 5$, where the eigenvalues of (13.8)–(13.9) are ordered according to $|\lambda_1|\leqq|\lambda_2|\leqq\cdots$. The exact eigenvalues of (13.3)–(13.5) are found by seeking solutions of these equations of the form $\psi(x,t)=\psi(x)\exp(\lambda t)$, $\zeta(x,t)=\zeta(x)\exp(\lambda t)$. It may be easily verified that the exact eigenvalues of (13.3)–(13.5) are given by $\lambda=-\mu^2$ with $\mu=n\pi$ *or* μ any nonzero solution of the transcendental equation $\tan\mu=\mu$. The exact values of λ_1 and λ_5 are also listed in Table 13.1. Evidently, even though (13.8)–(13.9) is unstable as $N\to\infty$, it does a good job of reproducing the low-n modes; approximately $\sqrt{|\lambda_n|}$ Chebyshev polynomials are required to resolve the mode with eigenvalue λ_n. Thus, this version of the tau method may be useful for eigenvalue calculations even though it is unconditionally unstable for the initial-value problem (13.3)–(13.5) (as evidenced by the spurious unstable modes with eigenvalues as large as $\lambda_{\max}$).

TABLE 13.1

Eigenvalues of the tau approximation (13.8)–(13.9) *to* (13.6)–(13.7). *The N-4 eigenvalues are ordered so that* $|\lambda_1|\leqq|\lambda_2|\leqq\cdots\leqq|\lambda_{N-4}|$. *All the eigenvalues are real. The largest positive eigenvalue is* $\lambda_{\max}=\lambda_{N-4}$.

N	λ_1	λ_5	$\lambda_{\max}$
10	−9.8696598	−189.63800	4,272.
15	−9.8696044	−89.54550	29,439.
20	−9.8696044	−88.86244	111,226.
25	−9.8696044	−88.86244	294,697.
30	−9.8696044	−88.86244	652,722.
35	−9.8696044	−88.86244	1,255,298.
40	−9.8696044	−88.86244	2,215,880.
Exact	−9.8696044	−88.86244	

The tau method behaves similarly when applied to more difficult problems, like the Orr–Sommerfeld equation for linear stability analysis of incompressible plane-parallel shear flows. Low modes are given accurately by the analogue of

(13.8)–(13.9) (see Orszag (1971d)), but there appear spurious unstable modes with large growth rates. Similar spurious unstable modes appear in finite-difference solution of the Orr–Sommerfeld equation (see Gary and Helgason (1970)).

There is a simple method to avoid the spurious unstable modes encountered by (13.8)–(13.9). The technique to be described below also eliminates the spurious unstable modes encountered in solution of the Orr–Sommerfeld equation. The idea is simply not to combine (13.3)–(13.4) into (13.6). Rather, we expand $\zeta(x, t)$ as in

(13.10) $$\zeta(x, t) = \sum_{n=0}^{N} b_n(t) T_n(x)$$

and solve

(13.11) $$\frac{db_n}{dt} = \nu b_n^{(2)}, \qquad 0 \leqq n \leqq N-2,$$

(13.12) $$b_n = a_n^2, \qquad 0 \leqq n \leqq N-2,$$

in addition to (13.9). Here we have dropped two equations from the Chebyshev modal equations that result from (13.3)–(13.4). The logic of this modification of the tau method is as follows. Application of (13.8) for $0 \leqq x \leqq N-4$ is equivalent to application of (13.12) for $0 \leqq n \leqq N$ together with (13.11) for $0 \leqq n \leqq N-4$. On physical grounds, we may expect that this procedure will lead to instability because the boundary conditions $\psi = 0$ at $x = \pm 1$ should be imposed on (13.4) not (13.3), while the boundary conditions $\psi_x = 0$ at $x = \pm 1$ should be imposed on (13.3) only when $\nu > 0$. On the other hand, when the system is truncated as in (13.11)–(13.12), each of the dynamical equations can play their proper role in adjusting the boundary conditions: the boundary conditions $\psi = 0$ are imposed on (13.12) while the boundary conditions $\psi_x = 0$ are imposed on (13.11).

We shall now prove that (13.11)–(13.12) is stable for the special case in which N is even with $a_{2n+1} = b_{2n+1} = 0$ for all n, $t \geqq 0$. In this case, $\psi(x, t)$ and $\zeta(x, t)$ are even functions of x. To begin, we observe that (13.11) is equivalent to

$$\frac{\partial \zeta}{\partial t} = \nu \frac{\partial^2 \zeta}{\partial x^2} + b'_N T_N(x), \qquad -1 \leqq x \leqq 1, \quad t > 0,$$

while (13.12) is equivalent to

$$\zeta(x, t) = \frac{\partial^2 \psi}{\partial x^2} + b_N T_N(x).$$

Therefore,

$$\frac{\partial^3}{\partial t\, \partial x^2} \psi = \nu \frac{\partial^4 \psi}{\partial x^4} + b_N T''_N.$$

Since ψ is an even function of x, it follows by integration with respect to x that

(13.13) $$\frac{\partial^2 \psi}{\partial x\, \partial t} = \nu \frac{\partial^3 \psi}{\partial x^3} + b_N T'_N.$$

Also, since $\psi(x, t)$ is a polynomial of degree N that satisfies $\psi_x(\pm 1, t) = 0$, integration by parts gives

$$\int_{-1}^{1} \psi_x T_N'(1-x^2)^{-1/2}\, dx = -\int_{-1}^{1} [\psi_{xx} + x\psi_x/(1-x^2)] T_N (1-x^2)^{-1/2}\, dx = 0$$

since ψ_{xx} and $x\psi_x/(1-x^2)$ are polynomials of degree $N-2$ so they must be orthogonal to $T_N(x)$. Therefore, taking the Chebyshev inner product of (13.13) and $\psi_x(x, t)$, we obtain

$$\frac{\partial}{\partial t}\int_{-1}^{1} \psi_x^2 (1-x^2)^{-1/2}\, dx = 2\nu \int_{-1}^{1} \psi_x \psi_{xxx} (1-x^2)^{-1/2}\, dx \leqq 0, \tag{13.14}$$

where the last inequality is established using the inequality derived in Example 7.1(v):

$$\int_{-1}^{1} u u_{xx} (1-x^2)^{-1/2}\, dx \leqq 0,$$

if $u(x)$ is a polynomial of degree N satisfying $u(\pm 1) = 0$. The energy bound (13.14) proves stability of the tau approximation (13.11)–(13.12).

Finally, let us discuss methods for the solution of the primitive equations (13.1) using Chebyshev–tau approximations. A one-dimensional model that embodies the essential features of (13.1) is obtained by solving (13.1) within the slab $-1 \leqq x \leqq 1$, $-\infty \leqq y \leqq \infty$, with an assumed solution of the form

$$\mathbf{v} = (u(x, t)\, e^{iky}, v(x, t)\, e^{iky}), \qquad p = p(x, t)\, e^{iky}$$

for some real wave number k. Let the Chebyshev expansion coefficients of $u(x, t)$, $v(x, t)$, $p(x, t)$ be denoted as $u_n(t)$, $v_n(t)$, $p_n(t)$ $(0 \leqq n \leqq N)$, respectively. Then an unconditionally stable, implicit fractional step method for the solution of (13.1) with a forcing term $(f(x, t)\, e^{iky}, g(x, t)\, e^{iky})$ added to the right side is

$$\bar{u}_n = u_n(t) + \Delta t\,[-p_n^{(1)} + f_n(t)], \qquad 0 \leqq n \leqq N-2, \tag{13.15}$$

$$\bar{v}_n = v_n(t) + \Delta t[-ikp_n + g_n(t)], \qquad 0 \leqq n \leqq N, \tag{13.16}$$

$$\bar{u}_n^{(1)} + ik\bar{v}_n = 0, \qquad 0 \leqq n \leqq N, \tag{13.17}$$

$$\sum_{n=0}^{N} \bar{u}_n = \sum_{n=0}^{N} (-1)^n \bar{u}_n = 0, \qquad 0 \leqq n \leqq N, \tag{13.18}$$

$$u_n(t+\Delta t) - \nu\, \Delta t\, u_n^{(2)}(t+\Delta t) = \bar{u}_n, \qquad 0 \leqq n \leqq N-2, \tag{13.19}$$

$$v_n(t+\Delta t) - \nu\, \Delta t\, v_n^{(2)}(t+\Delta t) = \bar{v}_n, \qquad 0 \leqq n \leqq N-2, \tag{13.20}$$

$$\sum_{n=0}^{N} (\pm 1)^n u_n(t+\Delta t) = \sum_{n=0}^{N} (\pm 1)^n v_n(t+\Delta t) = 0. \tag{13.21}$$

Here we use the notation that, for example, $u_n^{(2)}$ represents the Chebyshev coefficients of $u_{xx}(x, t)$. The scheme (13.15)–(13.21) is based on backwards Euler time differencing; it is straightforward to generalize (13.15)–(13.21) to other more accurate time-differencing methods.

The fractional step (13.15)–(13.18) involves computation of the pressure field by imposition of the incompressibility condition (13.17). Only the boundary conditions $u(\pm 1, t) = 0$ are applied because this part of the time step is effectively inviscid so only the normal flow can be specified at the boundary. Thus, we drop (13.15) for $n = N-1, N$ in favor of the two boundary conditions (13.18). The fractional step (13.19)–(13.21) involves the viscous term in (13.1) so boundary conditions are applied on both the normal velocity component u and the tangential velocity component v. Accordingly, the tau method involves dropping (13.19)–(13.20) for $n = N-1, N$ in favor of these boundary conditions.

The system (13.15)–(13.21) is solved as follows. Multiplying (13.15) by ik and subtracting the result from the Chebyshev x-derivative of (13.16) gives

$$\bar{v}_n^{(1)} - ik\bar{u}_n = v_n^{(1)}(t) - iku_n(t) + \Delta t[g_n^{(1)}(t) - ikf_n(t)], \qquad 0 \leqq n \leqq N-2.$$

Substituting $\bar{v}_n = i\bar{u}_n^{(1)}/k$ from (13.17) gives

$$\bar{u}_n^{(2)} - k^2\bar{u}_n = u_n^{(2)}(t) - k^2u_n(t) + \Delta t[-ikg_n^{(1)}(t) - k^2f_n(t)], \qquad 0 \leqq n \leqq N-2. \tag{13.22}$$

Equation (13.22) with the boundary conditions (13.18) is of the same form as (13.19)–(13.20) with boundary conditions (13.21). These equations are best solved by the algorithm discussed at the end of § 10.

The stability analysis of (13.15)–(13.21) is as follows. The evolution of a perturbation is governed by (13.15)–(13.21) with $f_n = g_n = 0$ for all n. Therefore, the solution of (13.22) is $\bar{u}_n = u_n(t)$ for all n. The implicit scheme (13.19)–(13.21) is an unconditionally stable scheme for solution of the heat equation. This proves that (13.15)–(13.21) is unconditionally stable.

Implicit, unconditionally stable methods for solution of (13.1) that do not use fractional steps [as in (13.15)–)13.21)] can also be implemented. In a fully implicit scheme, we would solve

$$u_n(t+\Delta t) = u_n(t) + \Delta t[-p_n^{(1)} + \nu u_n^{(2)}(t+\Delta t) + f_n(t+\Delta t)], \qquad 0 \leqq n \leqq N-2, \tag{13.23}$$

$$v_n(t+\Delta t) = v_n(t) + \Delta t\,[-ikp_n + \nu v_n^{(2)}(t+\Delta t) + g_n(t+\Delta t)], \qquad 0 \leqq n \leqq N-2, \tag{13.24}$$

$$u_n^{(1)}(t+\Delta t) + ikv_n(t+\Delta t) = 0, \qquad 0 \leqq n \leqq N, \tag{13.25}$$

$$\sum_{n=0}^{N} (\pm 1)^n u_n(t+\Delta t) = \sum_{n=0}^{N} (\pm 1)^n v_n(t+\Delta t) = 0. \tag{13.26}$$

Substituting (13.25) into (13.24) and eliminating p_n between (13.23)–(13.24), we obtain

$$\begin{aligned} u_n^{(2)}(t+\Delta t)-k^2u_n(t+\Delta t)&=u_n^{(2)}(t)-k^2u_n(t)\\ &\quad+\Delta t[\nu u_n^{(4)}(t+\Delta t)-\nu k^2u_n^{(2)}(t+\Delta t)-ikg_n^{(1)}-k^2f_n]\\ &\quad+b_1+b_2(-1)^n \qquad 0\leqq n\leqq N-2, \end{aligned} \tag{13.27}$$

$$\sum(\pm1)^n u_n(t+\Delta t)=\sum(\pm1)^n n^2u_n(t+\Delta t)=0. \tag{13.28}$$

Here b_1 and b_2 are parameters determined by the condition that (13.27)–(13.28) have a solution.

There are two features of the full implicit scheme (13.23)–(13.28) that should be mentioned in comparison with the fractional step scheme (13.15)–(13.22). First, the solution of (13.27)–(13.28) involves solution of an essentially pentadiagonal matrix equation in contrast to (13.22) which is essentially tridiagonal. This may be a serious difficulty because the pentadiagonal system (13.27) is not as well conditioned as the tridiagonal system (13.22).

Second, the full implicit scheme avoids an ambiguity of the Navier–Stokes equations pointed out by Orszag and Israeli (1974, p. 299). When the boundary conditions $\mathbf{v}=0$ are applied to (13.1) (with a force term included), we obtain

$$\nabla p=\nu\nabla^2\mathbf{v}+\mathbf{f} \tag{13.29}$$

on the boundaries. Therefore, we can obtain boundary conditions on both $\partial p/\partial n$ and p, where n is normal to the boundary. However, the divergence of (13.1) gives

$$\nabla^2 p=\nabla\cdot\mathbf{f}, \tag{13.30}$$

so p is the solution of the Poisson equation (13.30) satisfying both the Dirichlet and Neumann boundary conditions (13.29). It seems at first that p is overspecified. In fact, p is not overspecified; the above argument reflects only the adjustment that $\mathbf{v}$ must undergo near the boundary in order to make both sets of boundary conditions on p mutually consistent.

This adjustment process is directly accounted for in the system (13.23)–(13.28) but is only indirectly accounted for in the fractional step method (13.15)–(13.22). Only the boundary condition (13.18) is applied while determining the pressure in (13.15)–(13.17). Nevertheless, it seems that the fractional step method adjusts itself from time step to time step so no serious errors are produced by neglecting the tangential components of (13.29).

The methods discussed in this section extend to give stable methods for solution of the nonlinear Navier–Stokes equations. For example, if the forcing term (f, g) in (13.15)–(13.16) is chosen to be the nonlinear terms of the Navier–Stokes equations, our analysis shows that stability of (13.15)–(13.21) is determined by stability restrictions on the nonlinear terms alone.

Note added in proof. While the scheme (13.15)–(13–21) is stable, it is not consistent with (13.1). The scheme (13.23)–(13.26) is stable and consistent.

SECTION 14

Miscellaneous Applications of Spectral Methods

In this section, we survey some special topics regarding spectral methods. Some of these topics are still under active investigation, so the results reported here are very incomplete.

Complicated geometries. There are two ways that spectral methods can be used to solve problems in complicated geometries without introducing basis functions that are special to the geometry and, therefore, unwieldy and inefficient to use. The two methods are mapping and patching.

Mapping involves transforming the complicated domain into a simpler one by means of a coordinate transformation. Spectral methods are then applied in the simple geometry using the techniques discussed in earlier sections. For example, if we wish to solve the heat equation

$$\frac{\partial}{\partial t}u(x, y, t) = \nabla^2 u(x, y, t) \tag{14.1}$$

in the two-dimensional domain

$$-1 \leqq x \leqq 1, \qquad -f(x) \leqq y \leqq f(x)$$

for some given function $f(x)$ with the boundary conditions that $u = 0$ on the boundary of the domain, we would proceed as follows. First, we make the coordinate transformation

$$z = y/f(x), \qquad -1 \leqq z \leqq 1, \tag{14.2}$$

and rewrite (14.1) as

$$\begin{gathered}\frac{\partial}{\partial t}u(x, z, t) = \left(\frac{\partial}{\partial x} - \frac{f'}{f}z\frac{\partial}{\partial z}\right)^2 u(x, z, t) + f^{-2}\frac{\partial^2}{\partial z^2}u(x, z, t),\\ -1 \leqq x \leqq 1, \quad -1 \leqq z \leqq 1.\end{gathered} \tag{14.3}$$

Then, we expand $u(x, z, t)$ in a double Chebyshev series and integrate (14.3). For this purpose, a hybrid numerical scheme is recommended. Unconditionally stable time differencing can be obtained by a semi-implicit method (see § 10). Here a simple diffusion operator is added and subtracted from (14.3). The simple diffusion operator is then evaluated using a tau method (because the tau method is simplest when no complicated nonlinearities or nonconstant coefficient terms are involved); the remaining nonconstant coefficient term in (14.3) is then evaluated explicitly using fast Fourier transforms and the collocation method. The result is an efficient and accurate method for solution of (14.1).

Techniques like those just described have been applied to a variety of problems with much success. If a convenient coordinate transformation is available, the mapping technique combined with appropriate spectral methods may be expected to be very useful.

The idea of patching is that if the geometry is the union of several simpler geometries (like an L-shaped region) then spectral approximations can be formulated in each of the simpler domains. These solutions are then patched across the boundaries by requiring that the solution (and an appropriate number of derivatives) be smooth. When this technique is applied together with the mapping technique discussed above, it is possible to devise spectral shock-fitting methods for the solution of compressible flow problems. Patching methods require much further investigation to judge their usefulness in practical problems.

Poisson's equation in two and higher dimensions. The Chebyshev tau equations for Poisson's equation $\nabla^2 u = f$ in the square $-1 \leqq x \leqq 1$, $-1 \leqq y \leqq 1$ are

$$u_{nm}^{(2,0)} + u_{nm}^{(0,2)} = f_{nm}, \qquad 0 \leqq n \leqq N-2, \quad 0 \leqq m \leqq M-2, \tag{14.4}$$

while the Dirichlet boundary conditions $u = 0$ are

$$\sum_{n=0}^{N} (\pm 1)^n u_m = 0, \qquad 0 \leqq m \leqq M, \tag{14.5}$$

$$\sum_{m=0}^{N} (\pm 1)^m u_{nm} = 0, \qquad 0 \leqq n \leqq N. \tag{14.6}$$

Here we expand $u(x, y)$ and $f(x, y)$ in the double Chebyshev series

$$\begin{Bmatrix} u(x, y) \\ f(x, y) \end{Bmatrix} = \sum_{n=0}^{N} \sum_{m=0}^{M} \begin{Bmatrix} u_{nm} \\ f_{nm} \end{Bmatrix} T_n(x) T_m(y) \tag{14.7}$$

and we denote the Chebyshev expansion coefficients of $\partial^{p+q} u / \partial x^p \, \partial y^q$ by $u_{nm}^{(p,q)}$. The $2N+2M+4$ boundary conditions are not all linearly independent; there exist four linear relations among them, namely

$$\sum_{n=0}^{N} \sum_{m=0}^{M} (\pm 1)^n (\pm 1)^m u_{nm} = 0. \tag{14.8}$$

Thus, (14.4)–(14.6) gives $(N+1)(M+1)$ equations for the $(N+1)(M+1)$ unknowns u_{nm} $(0 \leqq n \leqq N, 0 \leqq m \leqq M)$.

By using (10.10) (or (A.23)), the system (14.4)–(14.6) can be reduced to a block tridiagonal matrix equation modified by extra full rows corresponding to the boundary conditions (14.5)–(14.6). These equations can be solved by standard block tridiagonal algorithms in order $N^3 M$ or order NM^3 operations. If Poisson's equation must be solved several times with the same values of N and M but different functions $f(x, y)$, it is more efficient to apply alternative methods.

A method to solve (14.4)–(14.6) in order N^2M operations (with a preprocessing stage that requires order N^3 operations) is as follows. First, we find the $N-2$ eigenvalues λ_p and eigenvectors e_{np} ($p=0,\cdots,N-2$) of the equations

$$e_{np}^{(2)}=\lambda_p e_{np}, \qquad 0 \leqq n \leqq N-2,$$

$$\sum_{n=0}^{N}(\pm 1)^n e_{np}=0.$$

The eigenvalues λ_p are all negative as proved in Example 7.3(ii). Then we form the $(N+1)\times(N+1)$ matrix E whose elements are

$$E_{np}=e_{np}, \qquad 0\leqq n \leqq N, \quad 0 \leqq p \leqq N-2,$$

$$E_{n,N-1}=\delta_{n,0}, \qquad 0 \leqq n \leqq N,$$

$$E_{n,N}=\delta_{n,1}, \qquad 0 \leqq n \leqq N,$$

and compute the inverse matrix $D=E^{-1}$. Since the boundary conditions (14.5) are satisfied by u_{nm}, it follows that

$$u_{nm}=\sum_{p=0}^{N-2} e_{np}v_{pm} \tag{14.9}$$

for suitable v_{pm} for all n, m. Therefore, setting

$$g_{pm}=\sum_{n=0}^{N}(D)_{pn}f_{nm}, \qquad 0 \leqq p \leqq N-2, \quad 0 \leqq m \leqq M-2, \tag{14.10}$$

we see that (14.4)–(14.6) become

$$\lambda_p v_{pm}+v_{pm}^{(0,2)}=g_{pm}, \qquad 0 \leqq p \leqq N-2, \quad 0 \leqq m \leqq M-2, \tag{14.11}$$

$$\sum_{m=0}^{M}(\pm 1)^m v_{pm}=0, \qquad 0 \leqq p \leqq N-2. \tag{14.12}$$

Equations (14.11)–(14.12) may be solved efficiently (in order NM operations) for v_{pm} using the algorithm discussed at the end of § 10. Once v_{pm} is found, u_{nm} may be reconstructed from (14.9). The total operation count is order N^2M (from the two matrix multiplics (14.9)–(14.10)).

The solution of Poisson's equation by the Chebyshev series method outlined above is very competitive with finite-difference solution using fast Poisson solvers. Zang and Haidvogel (1977) present a number of comparisons of the Chebyshev methods and fast Poisson solvers.

There are two further complications that may arise in elliptic boundary-value problems. First, the elliptic equation may have nonconstant coefficients or may even be nonlinear. Here we recommend that spectral equations be solved using relaxation methods of the kind advocated by Concus and Golub (1973), in which the heart of the algorithm is the fast, efficient solution of Poisson-like equations.

Second, the geometry may be more complicated than a box. In this case, we recommend the implementation of capacitance matrix techniques (or equivalent Green's function techniques) in which the problem to be solved is imbedded in a simpler geometry, like a box (see Buzbee et al. (1971)). Again, the heart of the algorithm is the fast solution of Poisson's equation using (14.9)–(14.12).

Coordinate singularities. When spectral methods are applied to problems in cylindrical or spherical geometries, their formulation may require special care at the coordinate singularities. These "pole problems" have been extensively investigated (Orszag (1974), Tang (1977)). As a simple example of these effects, let us consider the computation of the eigenvalues of Bessel's equation using the Chebyshev tau method (Metcalfe (1974)). The problem is to find the eigenvalues and eigenfunctions $y(x)$ of

$$y''+\frac{1}{x}y'-\frac{n^2}{x^2}y=-\lambda y \tag{14.13}$$

subject to the conditions that $y(1)=0$ and that $y(x)$ be finite for $0\leqq x\leqq 1$. The exact eigenvalues are related to the zeros of the Bessel function J_n: $\lambda_p=j_{np}^2$ where $J_n(j_{np})=0$, $p=1, 2, \cdots$.

When n is even, the eigenfunctions of (14.13) are even functions of x; when n is odd, the eigenfunctions are odd. This fact suggests that we represent the solution to (14.13) in terms of series of even Chebyshev polynomials when n is even and odd polynomials when n is odd. Thus, for n odd we write

$$y(x)=\sum_{m=1}^{M} y_m T_{2m-1}(x). \tag{14.14}$$

TABLE 14.1

Smallest eigenvalue of (14.13) *with* $n=7$ *obtained using* (14.14) *and the Chebyshev tau method. M is the number of Chebyshev polynomials. The extra boundary condition* $y'(0)=0$ *is a pole constraint at the singular point* $x=0$ *of* (14.13).

M	λ_1 with $y(1)=0$	λ_1 with $y(1)=y'(0)=0$
10		124.001290649
14	169.111983340	122.895944051
18	126.557832251	122.907620295
22	122.991799598	122.907600279
26	122.908250800	122.907600204
Exact	122.907600204	122.907600204

In Table 14.1, we list numerical values for the smallest eigenvalue of (14.13) with $n=7$ using the series (14.14), the boundary condition $y(1)=0$, and the Chebyshev tau method. The convergence of this method, while very impressive as M increases, is slowed by the coordinate singularity of (14.13) at $x=0$. In general, series of the form (14.14) behave like x as $x\to 0$. In this case the terms y'/x and

y/x^2 are singular at $x=0$. The true eigenfunctions $J_7(j_{7p}x)$ behave like x^7 as $x \to 0$, as may easily be shown using Frobenius' method, so none of the terms of (14.13) are in fact singular for the exact eigenfunctions.

It is possible to improve the convergence of (14.14) by imposing additional "pole conditions", like $y'(0)=0$. When $y'(0)=0$ in the series (14.14), the terms of (14.13) are individually nonsingular. In Table 14.1, we also list numerical values of the smallest eigenvalue of (14.13) with $n=7$ and the two boundary conditions $y(1)=0$, $y'(0)=0$ applied. There is clearly a dramatic improvement in the rate of convergence. It is also possible to make the problem less sensitive to pole properties near the origin by first multiplying (14.13) by x^2 to eliminate explicitly singular terms and then applying the tau method. The results of the latter trick are essentially the same as applying the pole condition $y'(0)=0$ directly to (14.13).

If pole conditions are not properly applied, it is possible to degrade significantly the accuracy of spectral computations. It is even possible to induce strong instabilities that are absent when proper pole conditions are applied. These matters are discussed in detail by Orszag (1974) and Tang (1977).

Energy conservation. It was shown in § 2 (see footnote on p. 17) that if $(u, A(u))=0$ so the solution to the nonlinear equation $\partial u/\partial t = A(u)$ conserves energy $[\partial(u, u)/\partial t = 0]$, then the solution to any spectral approximation obtained by a self-adjoint projection operator also conserves energy. Some examples of this result are energy conservation by Galerkin approximations to the inviscid Navier–Stokes equations ((9.55) with $\nu=0$) using Fourier series with periodic or free-slip boundary conditions and Legendre polynomial series with rigid no-slip boundary conditions.

If the inviscid Navier–Stokes equations are rewritten in so-called rotational form as

$$\frac{\partial \mathbf{v}}{\partial t} = \mathbf{v} \times \boldsymbol{\omega} - \nabla \Pi + \nu \nabla^2 \mathbf{v}, \tag{14.15}$$

where $\boldsymbol{\omega} = \nabla \times \mathbf{v}$ is the vorticity and $\Pi = p + \frac{1}{2}v^2$ is the pressure head, then energy conservation holds when $\nu=0$ for certain collocation approximations. If the collocation points are $\mathbf{x}_i$ and $\mathbf{v}_i = \mathbf{v}(\mathbf{x}_i)$, then (14.15) gives

$$\frac{\partial}{\partial t} \sum_i v_i^2 = -\sum_i \mathbf{v}_i \cdot \nabla \Pi_i \tag{14.16}$$

when $\nu=0$. If the collocation projection operator is such that integration by parts is valid in the sense that

$$\sum_i \mathbf{v}_i \cdot \nabla \Pi_i = -\sum_i \Pi_i \nabla \cdot \mathbf{v}_i + \text{boundary terms}, \tag{14.17}$$

then $\nabla \cdot \mathbf{v} = 0$ and the boundary conditions imply that energy conservation holds. Thus, Fourier collocation approximation conserves energy when periodic or free-slip boundary conditions are applied to (14.15) (see Fox and Orszag (1973)).

SECTION 15

Survey of Spectral Methods and Applications

In this section, we give a brief survey of spectral methods and some of their recent applications. There are five important features of spectral methods that should be considered in their formulation and application. They are:

(i) *Rate of convergence.* If the solution to a problem is infinitely differentiable, then a properly designed spectral method has the property that errors go to zero faster than any finite power of the number of retained modes. In contrast, finite-difference and finite-element methods yield finite-order rates of convergence. The important consequence is that spectral methods can achieve high accuracy with little more resolution than is required to achieve moderate accuracy.

(ii) *Efficiency.* The development of fast transform methods permits spectral methods to be implemented with comparable efficiency to that of finite difference methods with the same number of independent degrees of freedom. However, since spectral methods typically require a factor of 2–5 fewer degrees of freedom in each space direction to achieve moderate accuracy (say, 5% error), the spectral computations can be considerably more effective. As the required accuracy increases, the attractiveness of spectral methods increases.

(iii) *Boundary conditions.* As shown in earlier sections of this monograph, the mathematical features of spectral methods follow very closely those of the partial differential equation being solved. Thus, the boundary conditions imposed on spectral approximations are normally the same as those imposed on the differential equation. In contrast, finite-difference methods of higher order than the differential equation require additional "boundary conditions". Many of the complications of finite-order finite-difference methods disappear with the infinite-order-accurate spectral methods.

Another aspect of the treatment of boundary conditions by spectral methods is their high resolution of boundary layers. If the solution to a problem has a boundary layer of thickness ε, then only about $1/\varepsilon^{1/2}$ polynomials (see (3.50)) need be retained to achieve high accuracy. In contrast, finite-difference methods using equally spaced grid points would require about $1/\varepsilon$ grid points to resolve such a boundary layer solution. Moreover, if a coordinate transformation is employed to improve the resolution of a boundary or internal layer of thickness ε, the errors of spectral methods are decreased faster than any finite power of ε as $\varepsilon \to 0$.

(iv) *Discontinuities.* Surprisingly, spectral methods do a better job of localizing errors than difference schemes and hence require considerably less local dissipation to smooth discontinuities.

(v) *Bootstrap estimation of accuracy.* It is often possible to estimate the accuracy of spectral computations by examination of the shape of the spectrum. Thus, in computations of three-dimensional incompressible flows at high Reynolds numbers, the mean-square vorticity spectrum must not increase abruptly at large wave numbers (small scales); if the vorticity spectrum decreases smoothly to 0 as wave number increases, it is safe to infer that the calculation is accurate. On the other hand, similar criteria for finite-difference methods can be very misleading.

Let us now survey some applications of spectral methods to incompressible fluid dynamics. We shall classify the method according to the boundary conditions and geometry.

(i) *Periodic boundary conditions in Cartesian coordinates.* Here Fourier series are appropriate. Spectral methods have been regularly used in three dimensions with $32\times32\times32$ modes and in two dimensions with 128×128 modes to simulate homogeneous turbulence. Most operational codes now use pseudospectral (collocation) methods because aliasing errors are usually small. The key fast transform methods are described in detail by Orszag (1971c).

More recently, more ambitious spectral codes have been developed. The KILOBOX code employs 1024×1024 Fourier modes in two dimensions while the CENTICUBE code uses up to $128\times128\times128$ modes in three dimensions. These high resolution codes are now being used to study fundamental questions regarding high Reynolds number turbulence, including the structure of inertial ranges.

(ii) *Rigid boundary conditions in Cartesian coordinates.* Here Chebyshev or Legendre polynomials should be employed. Typical applications to date include numerical studies of turbulent shear flows and boundary layer transition. Pseudospectral (collocation) methods are normally used, with Chebyshev polynomials particularly convenient because fast Fourier transform methods can be applied.

(iii) *Rigid boundary conditions in cylindrical geometry.* Here Chebyshev or Legendre polynomials should be used in radius, Fourier series in angle, and either Fourier or Chebyshev series in the axial direction (depending on boundary conditions). Some technical aspects of the implementation of Chebyshev series in radius, including pole conditions, are discussed by Orszag (1974). Applications to date include studies of transition in circular Couette flow and pipe Poiseuille flow. In particular, it should be emphasized that Chebyshev polynomial expansions are much better suited for serious numerical work than the apparently more natural choice of Bessel function expansions in radius. There are two reasons: Chebyshev series converge faster to general functions regardless of their boundary conditions and Chebyshev–spectral methods can be implemented efficiently by fast transform methods.

(iv) *Problems in spherical geometry.* Here surface harmonic expansions, generalized Fourier series, and "associated" Chebyshev expansions all have attractive features. A detailed discussion of these methods is outside the scope of this monograph, but roughly speaking generalized Fourier series permit the most efficient transform methods to be developed followed by associated Chebyshev expansions and then surface harmonic expansions but surface harmonic expansions are best with regard to the pole problem. A variety of applications of these methods to global atmospheric flows have been made.

(v) *Semi-infinite or infinite geometry.* Here Chebyshev expansions are best if the domain can be mapped or truncated to a finite domain without serious error. There are two cases here: additional boundary conditions may or may not be required at "infinity". Here again the formulation of spectral methods follows closely the exact mathematics. If additional boundary conditions, like radiation or outflow boundary conditions, must be imposed on the truncated domain, then they should also be applied to the spectral method. On the other hand, if mapping without additional boundary conditions does not introduce a singularity in the exact equations, no boundary conditions at "infinity" are required in the spectral approximation.

APPENDIX

Properties of Chebyshev Polynomial Expansions

The Chebyshev polynomial of degree n, $T_n(x)$, is defined by

(A.1) $$T_n(\cos\theta)=\cos n\theta.$$

Thus, $T_0(x)=1$, $T_1(x)=x$, $T_2(x)=2x^2-1$, $T_3(x)=4x^3-3x$, $T_4(x)=8x^4-8x^2+1$, and so on. The Chebyshev polynomials are the solutions of the differential equation

(A.2) $$\sqrt{1-x^2}\frac{d}{dx}\sqrt{1-x^2}\frac{dT_n}{dx}+n^2T_n=0$$

that are bounded at $x=\pm1$. They satisfy the orthogonality relation

(A.3) $$\int_{-1}^{1} T_n(x)T_m(x)(1-x^2)^{-1/2}\,dx=\frac{\pi}{2}c_n\delta_{nm},$$

where $c_0=2$, $c_n=1$ for $n>0$. Some properties of Chebyshev polynomials are

(A.4) $$T_{n+1}(x)=2xT_n(x)-T_{n-1}(x),$$

(A.5) $$|T_n(x)|\leqq 1,\qquad |T_n'(x)|\leqq n^2,$$

(A.6) $$\frac{d^p}{dx^p}T_n(\pm1)=(\pm1)^{n+p}\prod_{k=0}^{p-1}(n^2-k^2)/(2k+1),$$

(A.7) $$\left|\frac{d^p}{dx^p}T_n(x)\right|=O(n^{2p}),\qquad n\to\infty,\quad p \text{ fixed},\quad |x|\leqq 1,$$

(A.8) $$\begin{gathered}T_n(+1)=(+1)^n,\quad T_{2n}(0)=(-1)^n,\quad T_{2n+1}(0)=0,\\ T_{2n}'(0)=0,\quad T_{2n+1}'(0)=(-1)^n n.\end{gathered}$$

The following formulae relate the expansion coefficients a_n in the series

$$f(x)=\sum_{n=0}^{\infty}a_nT_n(x),\qquad |x|\leqq 1,$$

to the expansion coefficients b_n of

$$Lf(x)=\sum_{n=0}^{\infty}b_nT_n(x),\qquad |x|\leqq 1,$$

for various linear operators L. We use the constants c_n and d_n defined by

$$c_0=2, \quad c_n=0\ (n<0), \quad c_n=1\ (n>0),$$
$$d_n=1\ (n\geqq 0), \qquad d_n=0\ (n<0).$$

Some formulae are:

(A.9) $$Lf=f'(x): \quad c_nb_n=2\sum_{\substack{p=n+1\\ p+n \text{ odd}}}^{\infty} pa_p,$$

(A.10) $$Lf=f''(x): \quad c_nb_n=\sum_{\substack{p=n+2\\ p+n \text{ even}}}^{\infty} p(p^2-n^2)a_p,$$

(A.11) $$Lf=xf(x): \quad b_n=\tfrac{1}{2}(c_{n-1}a_{n-1}+a_{n+1}),$$

(A.12) $$Lf=x^2f(x): \quad b_n=\tfrac{1}{4}\{c_{n-2}a_{n-2}+(c_n+c_{n-1})a_n+a_{n+2}\},$$

(A.13) $$\begin{aligned}Lf=x^4f(x): \quad b_n=\tfrac{1}{16}\{&c_{n-4}a_{n-4}+(c_{n-3}+c_{n-2}^2+2c_{n-2})a_{n-2}\\ &+(c_{n-2}+2c_{n-1}+c_{n-1}^2+c_n^2+c_n)a_n+(c_{n-1}+c_n+c_{n+1}+c_{n+2})a_{n+2}\\ &+a_{n+4}\},\end{aligned}$$

(A.14) $$Lf=\frac{f(x)-f(0)}{x}: \quad c_nb_n=2\sum_{\substack{p=n+1\\ p+n \text{ odd}}}^{\infty}(-1)^{(p-n-1)/2}a_p,$$

(A.15) $$Lf=\frac{f(x)-f(0)-f'(0)x}{x^2}: \quad c_nb_n=2\sum_{\substack{p=n+2\\ p+n \text{ even}}}^{\infty}(p-n)(-1)^{(p-n-2)/2}a_p,$$

(A.16) $$Lf=\frac{f'(x)-f'(0)}{x}: \quad c_nb_n=4\sum_{\substack{p=n+2\\ p-n\equiv 2 \bmod 4}}^{\infty} pa_p,$$

(A.17) $$\begin{aligned}Lf=\frac{f'(x)-f'(0)-f''(0)x}{x^2}: \quad c_nb_n=2&\sum_{\substack{p=n+3\\ p-n\equiv 3 \bmod 4}}^{\infty}(p-n+1)pa_p\\ &-\sum_{\substack{p=n+5\\ p-n\equiv 1 \bmod 4}}^{\infty}(p-n-1)pa_p,\end{aligned}$$

(A.18) $$Lf=xf'(x): \quad c_nb_n=na_n+2\sum_{\substack{p=n+2\\ p+n \text{ even}}}^{\infty} pa_p,$$

(A.19) $$\begin{aligned}Lf=x^2f'(x): \quad b_n=\tfrac{1}{2}\Big\{&(n-1)a_{n-1}+(n+1)(1+d_{n-1}+c_{n-1})a_{n+1}\\ &+4\sum_{\substack{p=n+3\\ p+n \text{ odd}}}^{\infty} pa_p\Big\},\end{aligned}$$

(A.20) $Lf = xf''(x)$: $c_n b_n = 2n(n+1)a_{n+1} + \sum_{\substack{p=n+3 \\ p+n \text{ odd}}}^{\infty} p(p^2-n^2-1)a_p,$

(A.21) $Lf = x^2 f''(x)$: $c_n b_n = n(n-1)a_n + \sum_{\substack{p=n+2 \\ p+n \text{ even}}}^{\infty} p(p^2-n^2-2)a_p,$

(A.22) $Lf = \dfrac{f(x)}{1-x^2}$, with $f(\pm 1) = 0$: $c_n b_n = -2 \sum_{\substack{p=n+2 \\ p+n \text{ even}}}^{\infty} (p-n)a_p.$

Also, if we expand $f^{(q)}(x)$ as in

$$\frac{d^q}{dx^q} f(x) = \sum_{n=0}^{\infty} a_n^{(q)} T_n(x),$$

then

(A.23) $$c_{n-1} a_{n-1}^{(q)} - a_{n+1}^{(q)} = 2n a_n^{(q-1)}, \quad n \geqq 1.$$

Properties of Legendre Polynomial Expansions

The Legendre polynomial of degree n, $P_n(x)$, is defined as the solution of the differential equation

(A.24) $$\frac{d}{dx}(1-x^2)\frac{dP_n(x)}{dx}+n(n+1)P_n(x)=0$$

that satisfies $P_n(1)=1$. Thus, $P_0(x)=1$, $P_1(x)=x$, $P_2(x)=\frac{1}{2}(3x^2-1)$, and so on. The Legendre polynomials satisfy the orthogonality relation

(A.25) $$\int_{-1}^{1} P_n(x)P_m(x)\,dx=\frac{2}{2n+1}\delta_{nm}.$$

Some properties of Legendre polynomials are

(A.26) $$(n+1)P_{n+1}(x)=(2n+1)xP_n(x)-nP_{n-1}(x),$$

(A.27) $$P_n(\pm 1)=(\pm 1)^n, \qquad P_n'(\pm 1)=(\pm 1)^{n-1}\tfrac{1}{2}n(n+1),$$

(A.28) $$|P_n(x)|\leqq 1, \qquad |P_n'(x)|\leqq \tfrac{1}{2}n(n+1), \qquad |x|\leqq 1.$$

(A.29) $$P_n(\cos\theta)=\left[\frac{2}{n\pi\sin\theta}\right]^{1/2}\sin\left[(n+\tfrac{1}{2})\theta+\frac{\pi}{4}\right]+O(n^{-3/2}),$$

$$n\to\infty;\quad \theta\neq 0,\quad \pi;\quad \theta \text{ fixed.}$$

If $f(x)$ is expanded in the Legendre series

$$f(x)=\sum_{n=0}^{\infty} a_nP_n(x)$$

then

$$a_n=\frac{2}{2n+1}\int_{-1}^{1} f(x)P_n(x)\,dx.$$

If L is a linear operator and

$$Lf(x)=\sum_{n=0}^{\infty} b_nP_n(x)$$

then the relation between b_n and a_n is as follows:

(A.30) $$Lf(x)=f'(x):\quad b_n=(2n+1)\sum_{\substack{p=n+1\\ p+n\text{ odd}}}^{\infty} a_p,$$

(A.31) $$Lf(x)=f''(x):\quad b_n=(n+\tfrac{1}{2})\sum_{\substack{p=n+2\\ p+n\text{ even}}}^{\infty}[p(p+1)-n(n+1)]a_p,$$

(A.32) $$Lf(x)=xf(x):\quad b_n=\frac{n}{2n-1}a_{n-1}+\frac{n+1}{2n+3}a_{n+1}.$$

References

Section 1: Courant and Hilbert (1953), Jeffreys and Jeffreys (1966), Kantorovich and Krylov (1964).

Section 2: Clenshaw (1957), Collatz (1960), Elliot (1961), Fox and Parker (1968), Kantorovich and Krylov (1964), Lanczos (1956), Orszag (1971a,b,c, 1972), Richtmyer and Morton (1967), Scraton (1965), Strang and Fix (1973), Wright (1964).

Section 3: Acton (1970), Clenshaw (1962), Courant and Hilbert (1953), Erdelyi et al. (1953), Fox and Parker (1968), Isaacson and Keller (1966), Lanczos (1956), Orszag and Israeli (1974), Rivlin (1969), Szegö (1959), Zygmund (1935).

Section 4: Godunov and Ryabenkii (1963), Kreiss (1962), Kreiss and Oliger (1973), Laptev (1975), Miller and Strang (1965), Richtmyer and Morton (1967).

Section 5: Barnett and Storey (1974), Bartels and Stewart (1972), Chorin et al. (1977), Lie and Engel (1888), Strang (1960), Richtmyer and Morton (1967).

Section 6: Fornberg (1975), Kreiss and Oliger (1973), Lanczos (1956, 1966), Lyness (1974), Orszag (1971c, 1972), Swartz and Wendroff (1969), Wengle and Seinfeld (1977).

Section 7: Price and Varga (1970), Richtmyer and Morton (1967), Swartz and Wendroff (1969).

Section 9: Gazdag (1976), Gottlieb and Gustaffson (1976), Kwizak and Robert (1971), Mesinger and Arakawa (1976), Richtmyer and Morton (1967), Roache (1972), Widlund (1966).

Section 10: Clenshaw (1955), Cooley and Tukey (1965), Cooley (1967), Metcalfe (1974), Orszag (1970, 1971c, 1974), Orszag and Israeli (1974), Patterson and Orszag (1971).

Section 11: Boris and Book (1976), Browning et al. (1973), Dubiner (1977), Lanzos (1966, 1973), Orszag (1971a), Orszag and Israeli (1974), Orszag and Jayne (1974).

Section 12: Haidvogel (1977), Roache (1972).

Section 13: Deville and Orszag (1977), Gary and Helgason (1970), Orszag (1971b,d, 1976a), Orszag and Israeli (1974).

Section 14: Buzbee et al. (1971), Concus and Golub (1973), Fox and Orszag (1973), Metcalfe (1974), Orszag (1974), Orszag and Israeli (1974), Tang (1977), Zang and Haidvogel (1977).

Section 15: Armstrong et al. (1970), Bourke (1972), Boyd (1977a,b), Clenshaw and Norton (1963), Collins and Dennis (1973), Deville and Orszag (1977), Dew and Scraton (1973), Eliasen et al. (1970), Fox & Orszag (1973), Francis (1972), Gazdag (1975), Grosch and Orszag (1977), Herbert (1976), Herring et al. (1974), Hoskins (1973), Israeli and Orszag (1976), Machenauer (1972), Merilees (1973), Merilees and Orszag (1977), Munson and Joseph (1971), Murdock (1977), Orszag (1970, 1974, 1976a,b, 1977), Orszag and Israeli (1974), Orszag (1977), Young (1974).

Appendix: Clenshaw (1955, 1957, 1962), Erdelyi et al. (1953), Fox and Parker (1968), Metcalfe (1974)

Bibliography

F. S. ACTON (1970), *Numerical Methods That Work*, Harper and Row, New York.

T. P. ARMSTRONG, R. C. HARDING, G. KNORR AND D. MONTGOMERY (1970), *Solution of Vlasov's equation by transform methods*, Methods in Computational Physics, Vol. 9, Academic Press, New York, p. 29.

S. BARNETT AND C. STOREY (1974), *Matrix Methods in Stability Theory*, Barnes and Noble, New York.

R. BARTELS AND G. STEWART (1972), *Solution of the matrix equation $AX+XB=C$*, Comm. Assoc. Comput. Mach., 15, p. 820.

W. BOURKE (1972), *An efficient, one-level, primitive-equation spectral model*, Mon. Weather Rev. 100, pp. 683–689.

J. P. BORIS AND D. L. BOOK (1976), *Flux-corrected transport. III: Minimal error FCT algorithms*, J. Computational Phys., 20, p. 397.

J. P. BOYD (1977a), *Pseudospectral methods for eigenvalue and nonseparable boundary value problems*, to be published.

——— (1977b), *The choice of spectral functions on a sphere: A comparison of Chebyshev, Fourier and associated Legendre expansions*, to be published.

G. BROWNING, H. O. KREISS AND J. OLIGER (1973), *Mesh refinement*, Math. Comp., 27, p. 29.

B. L. BUZBEE, F. W. DORR, J. A. GEORGE AND G. H. GOLUB (1971), *The direct solution of the discrete Poisson equation on irregular regions*, SIAM J. Numer Anal., 8, p. 722.

A. J. CHORIN, T. J. R. HUGHES, M. F. MCCRACKEN AND J. E. MARSDEN (1977), *Product formulas and numerical algorithms*, to be published.

C. W. CLENSHAW (1955), *A note on the summation of Chebyshev series*, Math. Tab. Wash., 9, p. 118.

——— (1957), *The numerical solution of linear differential equations in Chebyshev series*, Proc. Cambridge Philos. Soc., 53, p. 134.

——— (1962), *Chebyshev series for mathematical functions*, Math. Tab. Nat. Phys. Lab., 5, H.M. Stationery Office, London.

C. W. CLENSHAW AND H. J. NORTON (1963), *The solution of nonlinear ordinary differential equations in Chebyshev series*, Comput. J., 6, p. 88.

L. COLLATZ (1960), *The Numerical Treatment of Differential Equations*, Springer-Verlag, Berlin.

W. M. COLLINS AND S. C. R. DENNIS (1973), *Flow past an impulsively started circular cylinder*, J. Fluid Mech., 60, p. 105.

P. CONCUS AND G. H. GOLUB (1973), *Use of fast direct methods for the efficient numerical solution of nonseparable elliptic equations*, SIAM J. Numer. Anal., 10, p. 1103.

J. W. COOLEY (1967), *Applications of the fast Fourier transform method*, Proc. IBM Scientific Computing Symposium on Digital Simulation of Continuous Systems, IBM, p. 83.

J. W. COOLEY AND J. W. TUKEY (1965), *An algorithm for the machine computation of complex Fourier series*, Math. Comp., 19, p. 297.

R. COURANT AND D. HILBERT (1953), *Methods of Mathematical Physics*, Vol. 1, Interscience, New York.

M. DEVILLE AND S. A. ORSZAG (1977), to be published.

P. M. DEW AND R. E. SCRATON (1973), *The solution of the heat equation in two space variables using Chebyshev series*, J. Inst. Math. Appl., 11, p. 231.

M. DUBINER (1977), to be published.

E. ELIASEN, B. MACHENAUER AND E. RASMUSSEN (1970), *On a Numerical Method for Integration of the Hydrodynamical Equations with a Spectral Representation of the Horizontal Fields*, Rep. No. 2, Dept. of Meterology, Copenhagen Univ., Denmark, 35 pp.

D. ELLIOTT (1961), *A method for the numerical integration of the one-dimensional heat equation using Chebyshev series*, Proc. Cambridge Philos. Soc., 57, p. 823.

A. ERDÉLYI, W. MAGNUS, F. OBERHETTINGER AND F. G. TRICOMI (1953), *Higher Transcendental Functions*, vol. 2, McGraw-Hill, New York.

D. G. FOX AND S. A. ORSZAG (1973), *Pseudospectral approximation to two-dimensional turbulence*, J. Computational Phys., 11, p. 612.

L. FOX AND I. B. PARKER, (1968), *Chebyshev Polynomials in Numerical Analysis*, Oxford University Press, London.

B. FORNBERG (1975), *On a Fourier method for the integration of hyperbolic equations*, SIAM J. Numer. Anal., 12, 509.

P. E. FRANCIS (1972), *The possible use of Laguerre polynomials for representing the vertical structure of numerical models of the atmosphere*, Quart. J. Roy. Meteor. Soc., 98, p. 662.

J. GARY AND R. HELGASON (1970), *A matrix method for ordinary differential equations*, J. Computational Phys., 5, p. 169.

J. GAZDAG (1975), *Numerical solution of the Vlasov equation with the accurate space difference method*, Ibid., 19, p. 77.

——— (1976), *Time-differencing schemes and transform methods*, Ibid., 20, p. 196.

S. K. GODUNOV AND V. S. RYABENKII (1963), *Special criteria of stability of boundary-value problems for non-self-adjoint difference equations*, Uspekhi Mat. Nauk, 3, p. 211.

D. GOTTLIEB AND B. GUSTAFFSON (1976), *Generalized DuFort–Frankel methods for parabolic initial-boundary value problems*, SIAM J. Numer. Anal., 13, p. 129.

C. E. GROSCH AND S. A. ORSZAG (1977), *Numerical solution of problems in unbounded regions: Coordinate transforms*, J. Computational Phys., to appear.

D. HAIDVOGEL (1977), *Resolution of downstream boundary layers in the Chebyshev approximation to viscous flow problems*, to be published.

T. H. HERBERT (1976), *Periodic secondary solutions in a plane channel*, Proc. Fifth International Conference on Numerical Methods in Fluid Dynamics, A. I. van de Vooren and P. J. Zandbergen, eds., Springer-Verlag, Berlin, p. 235.

J. R. HERRING, S. A. ORSZAG, R. H. KRAICHNAN AND D. G. FOX (1974), *Decay of two-dimensional homogeneous isotropic turbulence*, J. Fluid Mech., 66, p. 417.

B. J. HOSKINS (1973), *Comments on "The possible use of Laguerre polynomials for representing the vertical structure of numerical models of the atmosphere*, Quart. J. Roy. Meteor. Soc., 99, p. 571.

E. ISAACSON AND H. B. KELLER (1966), *Analysis of Numerical Methods*, John Wiley, New York.

M. ISRAELI AND S. A. ORSZAG (1976), *Numerical investigation of viscous effects on trapped oscillations in a rotating fluid*, Proc. Fifth International Conference on Numerical Methods in Fluid Dynamics, A. I. van de Vooren and P. J. Zandbergen, eds., Springer-Verlag, Berlin, p. 241.

H. JEFFREYS AND B. S. JEFFREYS (1966), *Methods of Mathematical Physics*, Cambridge University Press, Cambridge.

L. V. KANTOROVICH AND V. I. KRYLOV (1964), *Approximate Methods of Higher Analysis*, P. Noordhoff, Groningen, the Netherlands.

H. O. KREISS (1962), *Über die Stabilitätsdefinition für Differenzengleichungen die partielle Differentialgleichungen approximieren*, Nordisk Tidskr. Informationsbehandling, 2, p. 153.

H. O. KREISS AND J. OLIGER (1973), *Methods for the Approximate Solution of Time Dependent Problems*, World Meteorological Organization/International Council of Scientific Unions, Geneva.

M. KWIZAK AND A. J. ROBERT (1971), *A semi-implicit scheme for grid point atmospheric models of the primitive equations*, Mon. Weather Rev., 99, p. 32.

C. LANCZOS (1956), *Applied Analysis*, Prentice-Hall, Englewood Cliffs, N J.

——— (1966), *Discourse on Fourier Series*, Hafner, New York.

——— (1973), *Legendre versus Chebyshev polynomials*, Topics in Numerical Analysis, J. J. Miller, ed., Academic Press, New York, p. 191.

G. LAPTEV (1975), *Conditions for the uniform well-posedness of the Cauchy problem for systems of equations*, Soviet Math. Dokl., 16, p. 65.

S. LIE AND F. ENGEL (1888), *Theorie der Transformationsgruppen*, Leipzig.

J. LYNESS (1974), *Computational techniques based on the Lanczos representation*, Math. Comp., 28, p. 81.

B. MACHENAUER AND E. RASMUSSEN (1972), *On the Integration of the Spectral Hydrodynamical Equations by a Transform Method*, Rep. No. 3, Dept. of Meteorology, Copenhagen University, Denmark, 44 pp.

P. E. MERILEES (1973), *Pseudo-spectral approximation applied to the shallow water equations on a sphere*, Atmosphere, 11, p. 13.

P. E. MERILEES AND S. A. ORSZAG (1977), to be published.

F. MESINGER AND A. ARAKAWA (1976), *Numerical Methods Used in Atmospheric Models*, vol. 1, World Meteorological Organization/International Council of Scientific Unions, Geneva.

R. W. METCALFE (1974), *Spectral methods for boundary value problems in fluid mechanics*, Ph.D thesis, M.I.T., Cambridge, MA.

J. J. H. MILLER AND W. G. STRANG (1965), *Matrix Theorems for Partial Differential and Difference Equations*, Stanford University Tech. Rep. CS28, Stanford, CA.

B. R. MUNSON AND D. D. JOSEPH (1971), *Viscous incompressible flow between concentric rotating spheres, Part 1, Basic Flow*, J. Fluid Mech., 29, p. 289.

J. W. MURDOCK (1977), *A numerical study of nonlinear effects on boundary layer stability*, AIAA Paper 77–127, AIAP, New York.

S. A. ORSZAG (1970), *Transform method for the calculation of vector-coupled sums: Application to the spectral form of the vorticity equation*, J. Atmos. Sci., 27, p. 890.

——— (1971a), *Numerical simulation of incompressible flows within simple boundaries: Accuracy*, J. Fluid Mech., 49, p. 75.

——— (1971b), *Galerkin approximations to flows with slabs, spheres, and cylinders*, Phys. Rev. Letters, 26 (1971), p. 1100.

——— (1971c), *Numerical simulation of incompressible flows within simple boundaries: Galerkin (spectral) representations*, Studies in Appl. Math., 50, p. 395.

——— (1971d), *Accurate solution of the Orr–Sommerfeld equation*, J. Fluid Mech., 50, p. 689.

——— (1972), *Comparison of pseudospectral and spectral approximation*, Studies in Appl. Math., 51, p. 253.

——— (1974), *Fourier series on spheres*, Mon. Weather Rev., 102, p. 56.

——— (1976a), *Turbulence and transition: A progress report*, Proc. Fifth International Conference on Numerical Methods in Fluid Dynamics, A. I. van de Vooren and P. J. Zandbergen, eds., Springer-Verlag, Berlin, p. 32.

——— (1976b), *Design of large hydrodynamics codes*, Computer Science and Scientific Computing, J. Ortega, ed., Academic Press, New York, p. 191.

——— (1977), *Numerical simulation of turbulent flows*, Handbook of Turbulence, Plenum Press, New York, p. 281.

S. A. ORSZAG AND M. ISRAELI (1974), *Numerical simulation of viscous incompressible flows*, Ann. Rev. Fluid Mech., 5, p. 281.

S. A. ORSZAG AND L. W. JAYNE (1974), *Local errors of approximate solutions to hyperbolic problems*, J. Computational Phys., 14, p. 93.

S. A. ORSZAG AND Y. H. PAO (1974), *Numerical computation of turbulent shear flows*, Advances in Geophysics, Vol. 18A, F. N. Frenkiel and R. E. Munn, eds., Academic Press, New York, p. 225.

S. A. ORSZAG AND G. S. PATTERSON (1972), *Numerical simulation of three-dimensional homogeneous isotropic turbulence*, Phys. Rev. Letters, 28, p. 76.

G. S. PATTERSON AND S. A. ORSZAG (1971), *Spectral calculations of isotropic turbulence: Efficient removal of aliasing interactions*, Phys. Fluids, 14, p. 2538.

H. S. PRICE AND R. S. VARGA (1970), *Error bounds for semidiscrete Galerkin approximations of parabolic problems with applications to petroleum reservoir mechanics*, Numerical Solution of Field Problems in Continuum Physics, American Mathematical Society, Providence, p. 74.

R. D. RICHTMYER AND K. W. MORTON (1967), *Difference Methods for Initial Value Problems*, Interscience, New York.

T. J. RIVLIN (1969), *An Introduction to the Approximation of Functions*, Blaisdell, Waltham, MA.

P. J. ROACHE (1972), *Computational Fluid Dynamics*, Hermosa, Albuquerque, N M.

H. SCHAMEL AND K. ELSÄSSER (1976), *The application of the spectral method to nonlinear wave propagation*, J. Computational Phys., 22, p. 501.

R. E. SCRATON (1965), *The solution of linear differential equations in Chebyshev series*, Comput. J., 8, p. 57.

W. G. STRANG (1960), *Difference methods for mixed boundary-value problems*, Duke Math. J., 27, p. 221.

G. STRANG AND G. J. FIX (1973), *An Analysis of the Finite Element Method*, Prentice-Hall, Englewood Cliffs, N J.

B. K. SWARTZ AND B. WENDROFF (1969), *Generalized finite difference schemes*, Math. Comp., 23, p. 37.

G. SZEGÖ (1959), *Orthogonal Polynomials*, American Mathematical Society, New York.

C. M. TANG(HUI) (1977), *Numerical simulation of fluid flow in spherical geometry and two-dimensional magnetohydrodynamic turbulence*, Ph.D. thesis, M. I. T., Cambridge, MA.

C. M. TANG AND S. A. ORSZAG (1977), *Two-dimensional turbulence on the surface of a sphere*, to be published.

H. WENGLE AND J. SEINFELD (1977), *Pseudo-spectral solution of atmospheric diffusion problems*, J. Computational Phys., to appear.

O. WIDLUND (1966), *Stability of parabolic difference schemes in the maximum norm*, Numer. Math., 8, p. 186.

K. WRIGHT (1964), *Chebyshev collocation methods for ordinary differential equations*, Comput. J., 6, p. 358.

R. E. YOUNG (1974), *Finite-amplitude thermal convection in a spherical shell*, J. Fluid Mech., 63, p. 695.

T. ZANG AND D. HAIDVOGEL (1977), to be published.

A. ZYGMUND (1935), *Trigonometrical Series*, Dover, New York.

Index

V

W